Two Manuals Combined into One Book

AF800593

Marine Physical Fitness Training and Test

plus

The Army Weight Control Program for all Personnel

Department of Defense

Pentagon
Publishing

CONTENTS

CHAPTER	
	INTRODUCTION
1	PHYSICAL CONDITIONING PROGRAM (PCP)
2	PHYSICAL FITNESS TEST (PFT)
3	BODY COMPOSITION PROGRAM (BCP)

APPENDIX	
A	GLOSSARY
B	DAILY 16 PROGRAM
C	REMEDIAL PHYSICAL CONDITIONING PROGRAM (RPCP)
D	BODY COMPOSITION PROGRAM (BCP) ASSIGNMENT PROCESS
E	AGE 46 AND OVER RISK FACTOR SCREENING CHECKLIST
F	PHYSICAL FITNESS TEST AND BODY COMPOSITION WORKSHEET
G	PHYSICAL FITNESS TEST SCORING TABLE
H	ALTITUDE COMPENSATION FOR 3.0 MILE PHYSICAL FITNESS TEST RUN CHART
I	HEIGHT/WEIGHT STANDARDS AND BODY FAT PERCENTAGE TABLES
J	PHYSICAL FITNESS TEST ENHANCEMENT TRAINING
K	LIGHT/LIMITED DUTY ASSIGNMENT PROCESS
L	PREGNANCY PHYSICAL FITNESS TEST AND BODY COMPOSITION REQUIREMENTS

INTRODUCTION

0001. PURPOSE. To promulgate policy and implementing instructions, concerning physical fitness and body composition in the Marine Corps.

0002. STATUS

1. The policies and procedures in this Manual apply to HQMC staff agencies and COs who are authorized to issue directives concerning this program.

2. The CMC (C 472TP) must authorize any deviation from the instructions in this Manual.

0003. RESPONSIBILITY. The CMC (C 472TP) is responsible for the accuracy, currency, modification, and distribution of this Manual. HQMC staffs and COs are responsible for the timely entry of changes and the physical maintenance of copies of this Manual.

0004. ALLOWANCES

1. Submit requests for changes in allowances of this Manual to the CMC (C 472TP). Submit requests per reference (b).

2. Request missing pages by requisitioning the basic Manual and pertinent changes per reference (b).

0005. ORGANIZATION

1. This Manual is organized into chapters identified by an Arabic numeral as listed in the overall contents.

2. Paragraph numbering is based on four digits. The first digit indicates the chapter; the next digit, the section; the final two digits the general major paragraph number; and the combinations which follow the decimal point, the subparagraph number; e.g., 3103.3a(2) refers to chapter 3, section 1, general major paragraph number 03, subparagraph 3a(2).

3. Pages are numbered in separate series by chapter number, with the chapter number preceding each page number; e.g., the fourth page of chapter 2 is shown as 2-4.

4. A glossary of terms is provided in appendix A to assist users with definitions of the numerous terms and acronyms throughout this Manual.

0006. CHANGES. Printed changes are posted per reference (c). Interfile changes in the basic Manual in consecutive order and complete the Record of Changes page.

CHAPTER 1

PHYSICAL CONDITIONING PROGRAM (PCP)

	PARAGRAPH	PAGE
SCOPE	1100	1-3
RESPONSIBILITY	1101	1-3
OBJECTIVES	1102	1-3
PHYSICAL CONDITIONING	1103	1-4
MARINE CORPS PCP REQUIREMENTS	1104	1-7
CO's ACTION	1105	1-7

CHAPTER 1

PHYSICAL CONDITIONING PROGRAM (PCP)

1100. SCOPE. Every Marine must be physically fit, regardless of age, grade, or duty assignment. Fitness is essential to the day-to-day effectiveness and combat readiness of the Marine Corps. Furthermore, physical fitness is an indispensable aspect of leadership. The habits of self-discipline required to gain and maintain a high level of physical fitness are inherent to the Marine Corps way of life and must be a part of the character of every Marine. Marines who are not physically fit can be a detriment to the readiness and combat efficiency of their unit. Accordingly, every Marine will engage in an effective PCP on a continuing and progressive basis.

1101. RESPONSIBILITY. In order to maintain continuity of standards, the procedures in this Manual will be used for developing and administering the COs conditioning, testing, and remedial programs, and to clarify Body Composition Program (BCP) policies (formerly known as the Weight Control Program). It is the responsibility of COs to provide a fair evaluation of Marines' physical fitness and body composition through the procedures outlined in this Manual. When necessary, an Appropriately Credentialed Health Care Provider (ACHCP), which will be referred to throughout this Manual as a Medical Officer, or "MO", will be required to conduct a thorough physical evaluation; an ACHCP is defined in the Glossary, appendix A of this Manual. This evaluation will include at a minimum, a Body Composition Evaluation and a determination of whether or not the weight and body fat gain is due to an underlying cause or associated disease.

1102. OBJECTIVES

1. Overall Program Focus. The Unit's PCP should focus on combat conditioning, health, fitness, and unit cohesion rather than preparation exclusively for the semi-annual Physical Fitness Test (PFT). The Program should reflect the following objectives:

a. To contribute to the health and well-being of every Marine through regular exercise, fitness testing, and health education.

b. To develop Marines who are physically capable of performing their duties in garrison and in combat.

c. To develop in Marines a reserve level of physical fitness and endurance that will enhance their chances of success in a combat situation.

d. To provide a medium for developing the individual Marine's self-confidence and desire to excel, thereby enhancing the unit's overall discipline, morale, esprit de corps, and combat readiness.

2. CO's Focus. COs should strive to achieve the primary objective of combat readiness by emphasizing combat conditioning at the unit level. The CO should ensure this objective is reflected in the training plans and schedules, affording the individual Marine an opportunity to achieve a competent level of combat fitness.

3. Command Physical Training (PT) Representative's Focus. The CO will assign in writing a Marine SNCO (normally from the Operations and Training Section), as the Command PT Representative. This Marine will be the commander's senior representative in the administration of Marine Corps policy related to physical conditioning, testing and remedial programs. Of particular emphasis, this Marine will be responsible for:

a. Conduct of the PFT. Responsible to the CO for selection and training of PFT monitors and for all matters related to the PFT (safety, course selection, recording, etc.).

b. Body Composition Evaluations. To ensure accuracy in body composition evaluations, the **only** personnel authorized to conduct these measurements are the Command PT Representative and his/her designate.

1103. PHYSICAL CONDITIONING. The definition of physical fitness is subjective to those defining it. The Marine Corps considers physical fitness to be the ability of a Marine to meet the physical demands of any combat or duty situation without undue fatigue. To achieve this state of physical fitness, physical conditioning should incorporate the components and principles listed below.

1. Components of Physical Conditioning. To capitalize on those components that can benefit conditioning efforts, the following categories of exercises should be included in both individual and unit PCPs:

a. Strength. Muscular strength refers to the ability of the muscular system to move the body through resistance. Many associate strength training with progressive resistive exercises using weights and machines. However, the ability of a Marine to effectively handle their own body weight should be a prerequisite before integrating strength training with machines into their program. This can be accomplished through the Daily 16 Program described in paragraph 1103.3a, and applying the principles listed in paragraph 1103.2. Strength training can be broadly separated into two categories, general and specific.

(1) General Strength Training. This type of training strengthens the muscular system by focusing on a full body workout for strength and size. In this type of training, the major muscle groups are exercised without a specific task or functional goal in mind. This type of strength training contributes to overall health.

(2) Specific Strength Training. This type of strength training is task specific. For example, Marines desiring to climb a rope better would do rope climb training wearing body armor, and focus their strength training on muscles involved in rope climbing. A company of Marines expecting to operate in hilly terrain would focus their strength training primarily on lower body strength.

b. Endurance. Two types of endurance conditioning are needed for a Marine to meet the physical demands of combat, aerobic and anaerobic.

(1) Aerobic Endurance. Aerobic activity, meaning "in the presence of oxygen," is categorized by physical demands that are sub-maximal (not an "all out" effort) and involve activity that is continuous in nature (lasting more than 3-5 minutes). Two examples are road marching and long distance running.

(2) Anaerobic Endurance. Anaerobic activity, meaning "without oxygen," is categorized by physical demands that are high intensity and of shorter (less than 2-3 minutes) duration. Examples are rope climbing, most forms of weight lifting, and running short, quick distances.

c. Mobility. Mobility conditioning is geared towards improving quality of movement. Quality of movement depends on the following:

(1) Posture
(2) Balance and stability
(3) Agility
(4) Coordination
(5) Power
(6) Speed
(7) Flexibility

2. Principles of Physical Conditioning. There are several different principles to consider when developing an effective PCP:

a. Progression. Conditioning programs must incorporate a systematic means to increase training load.

b. Regularity. To realize a conditioning effect, training programs must conduct physical conditioning sessions at least 3-5 times per week.

c. Overload. Only when the various systems of the body are overloaded will they become able to handle greater load.

d. Variety. Varying a program from time to time maintains interest and prevents staleness.

e. Recovery. Essential for allowing the systems overloaded during conditioning to adapt and become stronger.

f. Balance. Balanced conditioning programs ensure all the components of physical fitness conditioning (strength, endurance, and mobility) are properly addressed.

g. Specificity. Conditioning that is specific in nature yields specific gains. For example, stationary bike riding is of little value in improving running.

3. Specific PCPs. Units that strive to augment their PCP should use innovative combinations of the types of exercise defined below:

a. Daily 16 Program. The Daily 16 Program is a comprehensive series of warm-up, conditioning, and cool-down exercises replacing the former Daily 7 Program. This all-encompassing program can be incorporated into any unit aerobic or anaerobic conditioning session, or can be used as a conditioning session in of itself. Appendix B contains the Daily 16 Program in detail.

(1) Warmup Exercises. The Daily 16 warm-up exercises facilitate gradual distribution of blood flow to the muscles, preparing both the cardiovascular and musculoskeletal systems for the exercise session, by effectively targeting both the upper and lower body. The increased blood flow to the muscles produces a warming effect, increasing the elasticity of the muscles and connective tissue, which is believed to reduce injury.

(2) Conditioning Exercises. The Daily 16 conditioning exercises provide a total body workout through the proper execution of traditional calisthenics. Conditioning exercises can be used as a workout session in and of itself, or to augment the main fitness event (e.g., squad ability run or obstacle course).

(3) Cool-down Exercises. The Daily 16 cool-down exercises (which are basically the same as the warm-up exercises) allow the body to gradually return to the pre-exercise state.

b. Occupational Conditioning. This conditioning is comprised of general physical conditioning exercises that will develop and maintain strength, endurance, and the physical skills necessary to sustain a Marine during combat. Good examples of occupational conditioning that prepares Marines to successfully handle the demands of their particular MOS/billet are Physical Readiness Training (PRT), progressive load-bearing marches, martial arts training, dry net training, Military Operation in Urban Terrain (MOUT) training, combat water survival training, obstacle course, and confidence course.

c. Competitive Conditioning. Competitive conditioning activities consist of teams or individuals competing against an opponent to win. This includes a combination of sports and military skills designed to foster the unit's combat readiness, competitive spirit, and esprit de corps. Some good examples are speed-marches reaction courses, orienteering, water-can re-supply relays, and stretcher or fireman carry relays.

d. Remedial Physical Conditioning Program (RPCP). All Marines who fail the PFT, demonstrate poor physical performance in MOS-related skills, or who are assigned to the BCP will be assigned to a RPCP. Remedial physical conditioning is a process by which Marines are conditioned in a progressive manner to meet prescribed standards. The goal is to provide challenging conditioning sessions, using the spectrum of aerobic conditioning, resistance conditioning, and other related exercises that will afford the Marines an opportunity to rebuild themselves after a weight and body fat increase, injury/illness, pregnancy, or a period that lacked a structured fitness program. The philosophy of remedial conditioning is not punitive in nature, but is intended to encourage improvement in overall fitness or to supplement a BCP. Comprehensive guidance for the development of a command directed RPCP is contained in paragraph 2400 and appendix C of this Manual.

e. Alternate Aerobic Conditioning. In cases of lower limb injury or related medical conditions that prevent running or hiking, Marines may opt to perform other low impact activities to supplement the requirement for cardiovascular conditioning. Swimming, deep water running, cycling, cross-country skiing, stair stepping, and rowing are excellent examples of low impact, endurance exercises that provide good augmentation or substitution for running or hiking regimens. For example, the advantages of swimming include the use of all major muscle groups during exercise, reducing lower body stress due to partial body weight being supported by the water, and the absence of impact on bones and joints. Additionally, the body's position in the water increases the efficiency of the circulation back to the heart during exercise. Although these types of aerobic conditioning alternatives may not improve overall running ability (e.g., speed and endurance), it can enhance a current running or hiking regimen, while minimizing related injuries.

1104. MARINE CORPS PCP REQUIREMENTS

1. Individual Requirements. The PCP requires every Marine to:

a. Participate in a minimum of 3 hours of physical fitness conditioning per week; 5 hours is encouraged if the unit's operation tempo can accommodate such a program.

b. Be tested semi-annually - PFT. (Reserves tested annually.)

c. Obtain a minimum 3rd Class score on all semi-annual PFTs.

1105. CO's ACTIONS. The PCP must provide opportunities for development through supervised individual or group effort, and sustained by common sense leadership at all unit levels. In order to develop and maintain the desired level of fitness, COs will:

1. Schedule a minimum of 3 hours per week of mandatory individual or group physical conditioning sessions. COs are encouraged to use normal working hours to satisfy this training requirement. It is recommended that each conditioning session be a minimum of 60 minutes in duration.

2. Establish, implement, and maintain the PCP requirements as stated in this Manual.

3. Ensure all Marines participate in exercises or activities commensurate with their medical qualifications and limitations.

4. Ensure proper administration and tracking for all Marines in a light and limited duty status (e.g., Morning Reports).

5. Ensure the unit's PCP is not constructed and implemented solely to prepare for a PFT, but that it emphasizes physical conditioning for combat. This distinction should be reflected in the unit training schedules.

6. Conduct a semi-annual PFT for all Marines per the instructions contained in chapter 2 of this Manual.

7. Place and track all Marines who fail the PFT on a RPCP per the guidance set forth in paragraph 2400 and appendix C.

8. Immediately process Marines for a BCP assignment when identified as not in adherence with weight and body fat standards and do not meet Physical Performance Evaluation criteria.

9. Place and track all Marines assigned to a BCP on a RPCP for the duration of their 6-month BCP assignment.

10. Employ Semper Fit resources to assist in the execution of the PCP, RPCP, and BCP. Semper Fit Fitness Centers staffed by qualified professionals, and the Semper Fit website (www.usmc-mccs.org) provide expert guidance in the areas listed below. The fitness centers, available at most major Marine Corps installations, can provide the following services:

a. Fitness assessments and exercise prescriptions*;

b. Exercise orientation;

c. Group exercise opportunities;

d. Instructional skill development; and

e. Health promotion programs.

*Note: Personal fitness assessments conducted by Semper Fit personnel (e.g., body fat testing), while valuable, will not be considered official for BCP testing.

11. Ensure timely and accurate reporting of PFT results and BCP assignments on fitness reports per reference (d), and that correct Unit Diary and SRB entries are made, as appropriate, per references (e) and (f).

12. Ensure the Command retains all completed PFT/BCP files for one year.

CHAPTER 2

PHYSICAL FITNESS TEST (PFT)

	PARAGRAPH	PAGE
GENERAL INFORMATION	2000	2-3

SECTION 1: TESTING PROCEDURES

CONDUCT OF THE PFT	2100	2-5
EXECUTION OF THE PFT	2101	2-6

SECTION 2: STANDARDS

PFT STANDARDS	2200	2-9
ALTITUDE 3.0 MILE RUN	2201	2-10
PARTIAL PHYSICAL FITNESS TEST (PPFT)	2202	2-10
ADMINISTRATIVE PROCEDURES	2203	2-10

SECTION 3: MEDICAL CONSIDERATIONS

SCOPE	2300	2-15
MEB	2301	2-15
PEB.	2302	2-17
BODY COMPOSITION STANDARDS AND LIGHT OR TLD	2303	2-18
REENLISTMENT/PROMOTION	2304	2-18
PREGNANCY/POSTPARTUM	2305	2-18
MEDICAL WAIVERS/EXEMPTIONS	2306	2-19
AGE 46 AND OVER RISK FACTOR SCREENING	2307	2-20

SECTION 4: REMEDIAL PROCEDURES

	PARAGRAPH	PAGE
RPCP	2400	2-21

TABLES

2-1	MINIMUM ACCEPTABLE PERFORMANCE REQUIREMENTS FOR PFT/PPFT EVENTS	2-9
2-2	MINIMUM PFT CLASSIFICATION SCORES	2-9
2-3	ADMINISTRATIVE ACTIONS	2-13

CHAPTER 2

PHYSICAL FITNESS TEST (PFT)

2000. GENERAL INFORMATION

1. Purpose. While units are responsible for testing their Marines for combat fitness and readiness as it pertains to their unit's mission, the PFT provides an instrument that measures the collective level of physical fitness Marine Corps wide. It is a measurement of general fitness vice combat readiness and unit/MOS capability. The PFT consists of three events: male Marines will perform dead-hang pull-ups, abdominal crunches, and a 3.0 mile run; and female Marines will complete the flexed-arm hang, abdominal crunches, and a 3.0 mile run. These events are designed to test the strength and stamina of the upper body, midsection, and lower body, as well as the efficiency of the cardiovascular system.

2. Personnel to be Tested. Units will administer the PFT semi-annually (calendar year) for score to all active duty Marines, regardless of age, gender, grade, or duty assignment.

a. PFT Not Required. The condition when a Marine will not be required to conduct a semi-annual PFT:

(1) Personnel on duty in a combat zone; the semi-annual PFT may be waived until their return from the combat zone.

(2) Pregnant Marines medically waived from taking a PFT.

(3) Marines in a light duty or limited duty status, restricting them from participating in the PFT or Partial PFT (PPFT). (Appropriate documentation by a Medical Officer (MO), that assigns light duty or directs a Medical Evaluation Board (MEB) action, is mandatory.)

(4) Marines awaiting separation based on a Physical Evaluation Board (PEB). Reference paragraph 2302 for further details.

b. PFT Requirement for End of Active Service (EAS)/Retirement

(1) Marines are required to take the PFT during the semi-annual period preceding their EAS or retirement date, unless directed by their CO. This date is not the date that terminal leave begins.

For example, a Marine whose EAS or retirement date is September, but is commencing terminal leave in June, will be required to take the semi-annual PFT for the January-June period. This Marine is not required to take the semi-annual PFT for the July-December period.

(2) Retirement/EAS Final Physical Examination. Completion of the required final physical examination, regardless of when it is completed, does not exempt a Marine from taking the PFT. A Marine, who elects to complete their final physical examination 7-12 months prior to EAS or retirement, will still be required to take the semi-annual PFT for that period.

For example, a Marine who completes the final physical in March, but does not EAS or retire until October, will still be required to take the semi-annual PFT for the January-June period.

c. Selected Marine Corps Reserve (SMCR). Reserve Marines are required to conduct an annual PFT for score. Reserve Marines on Active Duty Special Work (ADSW) orders for 30 days or more, can conduct their annual PFT with that unit. All scores are good for 2 years (for cutting score purposes only), if training does not permit testing.

CHAPTER 2

PHYSICAL FITNESS TEST (PFT)

SECTION 1: TESTING PROCEDURES

2100. CONDUCT OF THE PFT

1. Uniform. The authorized uniform for the semi-annual PFT is the USMC established green-on-green T-shirt and shorts, socks, and running shoes. The green-on-green sweat suit may be used in winter climates. Gloves may be worn. Substitutions of the aforementioned uniform are at the discretion of the CO.

2. Sequence of Events. The sequence of PFT events will be left to the discretion of the CO. All PFT events will be conducted in a single session, not to exceed 2 hours in duration. Movement of Marines from one event to the next should allow adequate time to recover, stretch, and drink water.

3. Restriction of Events. The command will not impose restrictions on the performance of any event other than those specifically stated in this Manual. For example, a command will not mandate that Marines must use the overhand grip when executing pull-ups or flexed-arm hang.

4. Monitoring. The Command Physical Training (PT) Representative will monitor the PFT and maintain the PFT and Body Composition Worksheets. Marines will be kept advised of their progress as they are performing each event. The Command PT Representative or designate will call out each repetition for the pull-ups, and time-passed (in intervals) for the flexed-arm hang and abdominal crunch. The Command PT Representative (and additional monitors) will be posted along the 3.0 mile run course. Marines will receive their split time at the halfway mark and finishing time.

5. Logistics. Minimal special-type equipment is required to conduct a PFT.

a. Pull-Up Bar Requirements

(1) Diameter of the bar may range between 1-1¾ inches; athletic tape on the bar is authorized.

(2) The bar must be high enough to allow the tallest Marine's legs to hang straight without touching the ground, when arms are fully extended.

(3) Bench or stool may be used for shorter Marines to initially grasp the bar.

b. 3.0 mile Run Course. The run course must be measured for accuracy and set over reasonably level ground. Course selection criteria must include consideration of the following:

(1) Safety. Operational Risk Management (ORM) measures must be taken to ensure the course does not expose participants to undue risk.

(2) Performance. When possible, the course should be an "out and back" or wide loop course. The run course should not include numerous sharp turns, etc. that would force a participant to slow down excessively to remain on the course.

(3) Location. An indoor or outdoor track is permissible for the conduct of the run event. A determination as to whether the track is a yard or meter track must be made and the track measured to ensure accurate distance, prior to conducting the PFT. Run courses should not require a Marine to do more than 12 laps to cover 3.0 miles, (e.g., a course should not be shorter than 440-yard per lap). Running this event on a treadmill is not authorized.

c. Support Equipment & Personnel (Monitors)

(1) A timepiece (digital or stopwatch) that accurately measures time to the second is required for the flexed-arm hang, abdominal crunch, and 3.0 mile run.

(2) Safety equipment and personnel will be required as prescribed per local SOP.

(3) Adequate supervisory personnel to monitor each event and act as timekeepers.

d. PFT and Body Composition Worksheets. Appendix F contains the worksheet for recording PFT and body composition data. The following information will be recorded on the worksheet and be maintained by the Command PT Representative:

(1) The Marine's name, SSN, rank, age, gender, height, weight, percent body fat (if required), age 46 and Over Risk Factor Screening (if required), Physical Performance Evaluation qualifier (if required), event time(s)/total number of repetitions, event score, total score, and age-based PFT classification.

(2) Points for each event will be awarded according to the PFT Scoring Table located in appendix G.

(3) Minimum acceptable performance requirements by event (time and repetitions), and age group are listed in table 2-1 in this chapter.

(4) Minimum PFT classification scores are listed in table 2-2 in this chapter.

2101. EXECUTION OF THE PFT

1. Individual performance on each PFT event will be executed as follows:

a. Pull-up. The goal of the pull-up event is for the Marine to execute as many accurate and complete pull-ups before dropping off the bar. The procedures are:

(1) This is not a timed event.

(2) Sweatshirts will be removed during the conduct of the pull-up event in order to observe the lockout of the elbows with each repetition.

(3) Assistance to the bar with a step up, being lifted up, or jumping up is authorized. Any assistance up to the bar will not be used to continue into the first pull-up.

(4) The bar must be grasped with both palms facing either forward or to the rear.

(5) The correct starting position begins when the Marine's arms are fully extended beneath the bar, feet are free from touching the ground or any bar mounting assist, and the body is motionless.

(6) The Marine's legs may be positioned in a straight or bent position, but may not be raised above the waist.

(7) One repetition consists of raising the body with the arms until the chin is above the bar, and then lowering the body until the arms are fully extended; repeat the exercise. At no time during the execution of this event can a Marine rest his chin on the bar.

(8) The intent is to execute a vertical "dead hang" pull-up. A certain amount of inherent body movement will occur as the pull-up is executed. However, the intent is to avoid a pendulum-like motion that enhances the ability to execute the pull-up. Whipping, kicking, kipping of the body or legs, or any leg movement used to assist in the vertical progression of the pull-up is not authorized. If observed, the repetition will not count for score.

(9) A repetition will be counted when an accurate and complete pull-up is performed.

b. Flexed-Arm Hang. The goal of the flexed-arm hang event is for a Marine to hang (maintain elbow flexion) for as long as possible. The procedures are:

(1) This is a timed event.

(2) Sweatshirts will be removed during the conduct of the flexed-arm hang event in order to observe when the Marine has completely locked-out her elbows.

(3) Assistance to the bar with a step up, being lifted up, or jumping up to the start position is authorized.

(4) The bar must be grasped with both palms facing either forward or to the rear.

(5) The correct starting position begins when the Marine's arms are flexed at the elbow, the chin is held above the bar and not touching it, and the body is motionless. At no time during the execution of this event can a Marine rest her chin on the bar.

(6) Marines are authorized to drop down below the bar, however, some degree of elbow flexion must be maintained with both arms. Once a Marine's arms are fully extended or the Marine drops off the bar, the clock will stop.

c. Abdominal Crunch. The goal of the abdominal crunch event is for a Marine to execute as many proper and complete crunches within the prescribed time limit. The procedures are:

(1) 2-minute time limit.

(2) On a flat surface, Marines will lie flat on their back with shoulder blades touching the deck, knees will be bent, and both feet will be flat on the deck.

(3) The arms will be folded across the chest or rib cage with no gap existing between the arms and chest/rib cage. Both arms must remain in constant contact with chest/rib cage throughout the exercise. A single repetition consists of raising the upper body from the starting position until both forearms or elbows simultaneously touch the thighs, and then returning to the starting position with the shoulder blades touching the deck.

(4) The buttocks will remain in constant contact with the deck throughout the event. No arching of the lower back or lifting the buttocks is permitted.

(5) An assistant may be used to hold a Marine's legs or feet, at or below the knees in whatever manner that is most comfortable for the Marine. Kneeling or sitting on the Marine's feet is permitted.

(6) A repetition will be counted when an accurate and complete abdominal crunch is performed.

d. 3.0 Mile Run. The goal is for a Marine to complete the measured course as quickly as possible. The procedures are:

(1) This is a timed event.

(2) On the command to start, the two Marines monitoring the event will start their watches simultaneously when the last Marine passes the starting point. The Command PT Representative will remain at the start/finish and the other monitor will take the safety vehicle (with communication capabilities) to the halfway point. Monitors will call out the split or finishing time as appropriate, as each Marine passes.

2. The PFT scoring table is contained in appendix G.

CHAPTER 2

PHYSICAL FITNESS TEST (PFT)

SECTION 2: STANDARDS

2200. PFT STANDARDS

1. PFT Performance Requirements. To successfully pass the PFT, Marines must complete the minimum acceptable performance requirements in each event and achieve an overall combined score for their age group as shown in table 2-1. Marines must be made aware that the minimum performance (points) in each event alone, will not total the points required for a passing score. Additional points must be earned in at least one event in order to achieve a 3rd Class PFT or better, per age group. Failure to meet the minimum requirements in any event constitutes a failure of the entire test, regardless of the total number of points earned for all three events. Marines on light or limited duty will complete the two events that they are medically qualified to participate in, and will receive credit for a PPFT. Paragraph 2202 further details the PPFT requirements.

AGE	PULL-UPS/ FLEXED-ARM	ABDOMINAL CRUNCHES	3.0 MILE RUN (MIN)	TOTAL POINTS	MIN SCORE	ADDNTL' POINTS NEEDED
17-26	3/15(SEC)	50	28(m) 31(f)	105	135	30
27-39	3/15	45	29(m) 32(f)	94	110	16
40-45	3/15	45	30(m) 33(f)	88	88	0
46+	3/15	40	33(m) 36(f)	65	65	0

Table 2-1.--Minimum Acceptable Performance Requirements for PFT/PPFT Events.

2. Individual event scores for both males and females can be viewed in appendix G.

3. PFT Classification Scores. Table 2-2 shows the minimum score required, per age group, to earn each PFT classification score. Marines should be encouraged to continually strive to perform their best and not merely accept minimum performance.

	AGE GROUPS			
PFT CLASS	17-26	27-39	40-45	46+
1st	225	200	175	150
2nd	175	150	125	100
3rd	135	110	88	65

Table 2-2.--Minimum PFT Classification Scores.

4. PFT scores will be entered into the MCTFS and on FitRpts per paragraph 2203 in this chapter.

2201. ALTITUDE 3.0 MILE RUN. Marines permanently assigned or on TAD to locations at or above 4500 feet above sea level will have adjusted run times. The altitude 3.0 mile run point values are contained in appendix H. Commands are directed to provide newly joined Marines a 30-day acclimation period prior to conducting a semi-annual PFT. Marines scheduled to report to a command at altitude in June or December will complete their PFTs prior to reporting.

> CH 1 2202. PARTIAL PHYSICAL FITNESS TEST (PPFT). Marines who are medically excused from performing a particular event of the PFT as certified by a MO (e.g., light or TLD), are required to perform a PPFT. The PPFT for score requires that Marines complete two of the three PFT events. For the event not conducted, the event score from the previous semi-annual PFT will be used. Event times/repetitions will be awarded points the same way the PFT awards points; reference appendix G to view the scoring table. As with a PFT, failure to meet the minimum requirements in either event constitutes a test failure. Reference table 2-1 to view the minimum acceptable performance requirements for each event.

1. Reporting

a. The PFT and Body Composition Worksheet will indicate the event not taken and whether a light duty, TLD, or PLD status was the reason for the PPFT.

>CH 1 b. Report PPFT performance (pass/fail) for Sgt and above in accordance with reference (e).

c. LCpls and Cpls requiring a composite score will use the score (points) received from the last semi-annual PFT for the event missed.

d. PPFT scores will be entered on the Unit Diary and FitRpt per paragraph 2203.

2. Post-Light or Limited Duty PFT. Marines have the option to either keep the PPFT score taken for the current semi-annual period or to conduct a PFT (e.g., for promotion purposes) after being returned to full duty. The Marine will be given a minimum of 30 days and no more than 90 days to complete the PFT. This score will replace the PPFT taken during the waiver period. This is the only time a retest score will be used to replace a previous entry (partial) reported in MCTFS; this is not considered a remedial score.

3. Consecutive PPFTs. Marines who take PPFTs for two consecutive semi-annual periods, for the same injury or illness, will be referred to a MO and processed for a medical board. A MEB, PEB, or both will be conducted in order to determine the Marine's fitness for duty, per reference (i).

2203. ADMINISTRATIVE PROCEDURES

1. PFT/PPFT Recording

a. Unit Diary. Only when the command declares that the PFT scheduled is for the semi-annual requirement, will scores be reported by a Unit Diary entry and entered in the MCTFS, per reference (e). Only one PFT score will be reported on the Unit Diary per semi-annual period. No retest or remedial PFT scores will be recorded on the Unit Diary. PFTs taken for inventory purposes, such as a Marine reporting into a new unit or school, or taken for screening checklist purposes, will not be reported on the Unit Diary. In addition, PFT scores conducted for unit PT or for Remedial Physical Conditioning Program (RPCP) testing may not be used (or replace a score for) the semi-annual requirement.

b. FitRpt. All PFT score reporting will be done per reference (d), paragraph 4003.8b.

(1) Semi-annual PFT scores will be recorded in section A, "Administrative Information," under "Special Information" in block 8b.

(2) Mandatory comments will be made in section I, "Directed & Additional Comments," for the following circumstances:

(a) Superior Performance. Score of 285 and above.

(b) Not Medically Qualified (NMED). Limited duty status. Refer to paragraph 2300.3 for further details.

(c) PPFT. Sgts and above that have been granted a medical excuse by a MO, and who complete and pass a PPFT will have PART entered in block 8b, section A of their FitRpt. A brief explanation of the necessity for the PPFT, (e.g., TLD or PLD must be included in section I. A failed PPFT will be reported the same as for the PFT.

(d) PFT/PPFT Failure. A PFT/PPFT failure requires that Sgts and above receive an adverse FitRpt at the end of the next reporting period. In the case where a Marine does not have a reporting period until after the next semi-annual PFT, the previous PFT failure must be recorded on that FitRpt report, regardless of the performance on the more recent one.

(e) Required but Did Not Take (RDNT). A RDNT requires Sgts and above receive an adverse FitRpt. RDNT will be recorded in block 8b, section A and an explanation must be included in section I of the FitRpt.

(3) Height and weight measurements (and percent body fat, when necessary) will be transcribed from the PFT and Body Composition Worksheet for that reporting period.

c. Proficiency/Conduct (PRO/CON) Marks. Any Marine ranked Cpl or below, will receive commensurate PRO/CON marks for superior performance, PFT failure, or RDNT per reference (g), para 4005.

d. Page 11 SRB/OQR Entries. Page 11 entries are not required for Marines who fail a PFT/PPFT or for when a PFT/PPFT is RDNT.

2. PFT/PPFT Failures

a. Promotion/Reenlistment/Special Schools. Marines may be denied promotion, reenlistment, and/or special schools until they pass a PFT and fulfill all RPCP requirements. See paragraphs 1204 and 6009.2.e of reference (j).

b. Administrative Separation. Marines may be subject to administrative separation, per chapter 6 of reference (h), paragraphs 6105 and 6206, for the following circumstances:

(1) Unsatisfactory Performance. Repetitive PFT failures, due to lack of physical conditioning, displays poor self-discipline in a Marine's responsibility to maintain fitness standards.

(2) Medical Condition. Repetitive PPFT failures due to a medical condition.

3. Summary of Administrative Actions. Table 2-3 lists the administrative actions required for Marines who fail the PFT/PPFT.

ADMINISTRATIVE ACTION	NUMBER OF FAILURES	
	PFT 1st	PFT 2nd+
Mandatory RPCP	YES	YES
Page 11 SRB/OQR Mandatory Entry	NO	NO*
Adverse FitRpt or Substandard PRO/CON Marks	YES	YES
Eligible for Promotion	NO	NO
Eligible for PCS Transfer:	YES	YES
Process for Administrative Separation	NO	YES
May Be Denied Re-enlistment	YES	YES
May Be Denied Special Schools	YES	YES
May Be Denied Recommendation for Promotion	YES	YES

Table 2-3.--Administrative Actions.

*[Page 11 entries are required when Marines have failed consecutive PFTs or PPFTs, and when process for administrative separation is anticipated. See paragraph 6206, reference (h).]

CHAPTER 2

PHYSICAL FITNESS TEST (PFT)

SECTION 3: MEDICAL CONSIDERATIONS

2300. SCOPE. When a Marine is unable to take or pass the PFT due to a physical or medical condition, then a medical evaluation by a MO must be conducted, in order to excuse the Marine from participating in a particular event or the entire test. If the condition becomes extended or considered permanent in nature, then a determination by a formal MEB, a PEB, or both must be conducted to determine the Marine's medical qualification for continued active service. Marines evaluated for or assigned to a light duty or limited duty status are still required to perform those events of the PFT that are not affected by the condition, unless otherwise specified by the MO. A Marine may not be repeatedly excused from participation in the PFT/PPFT without a MEB action taken.

1. Light Duty. The status a MO assigns to a Marine when it is determined that a medical condition exists that temporarily interferes with the performance of their duty, but the Marine is expected to return to full duty (RTFD). A MO may recommend up to 30 days light duty. A Marine that is reevaluated and not returned to a full duty status in 30 days will be placed on medical hold, and a MEB must be initiated and completed within the following 30 days. Reference paragraph 2301 for further details on the MEB.

2. TLD. The status a MO assigns to a Marine when it is determined that a medical condition exists that requires an extended period of time before they can RTFD.

a. Marines, who are on TLD and are medically exempt from two consecutive PFTs or PPFTs for the same injury or illness, will be referred to a MO and processed for a MEB, a PEB, or both to determine fitness for duty.

b. A MEB must be conducted in order to assign a Marine to a TLD status. A Marine can be assigned to a TLD status for no more than 1-8 months by a medical treatment facility (MTF).

c. A PEB must be conducted in order to grant a Marine an extension of TLD, to be assigned a PLD status, or to be processed for separation.

3. Post-Light or Limited Duty PFT. Marines removed from light or TLD status by a MO, who did not take a PFT/PPFT for the semi-annual requirement, will be administered a PFT. COs should be attentive to the fact that a Marine returned to full duty status will need to train at a moderate pace to return to the level of fitness experienced prior to the injury/illness. As a result, a Marine will be given no less than 30 days and no more than 90 days preparation after being returned to a full duty status, to complete the PFT. If the grace period crosses over into the next semi-annual interval, the PFT score will satisfy the later requirement and PFT MED (medically excused) will be entered into MCTFS for the earlier testing period.

2301. MEB. A MEB is a group of MO's at a local MTF convened for the purpose of identifying the physical and mental qualifications of referred Marines to RTFD. The MEB is charged with determining whether a Marine's limitation(s) preclude their RTFD within a reasonable period of time or at all. A MEB will evaluate and report on the diagnosis, prognosis, possible RTFD, plan for further treatment, and make medical recommendations for disposition. Reference appendix K of this Manual for further details on when a MEB is required.

1. Case Referrals. A MEB case referral will be made on Marines not returned to a full duty status after the maximum 30 days of light duty, or medically waived from completing a PFT or PPFT for two consecutive semi-annual periods for the same medical condition. A MEB may recommend:

a. Returning the Marine to full duty.

b. A period of TLD if the initial light duty period of 30 days has not rectified the problem.

c. Placing the Marine on medical hold. Medical hold is a temporary status, not to exceed 30 days, which is used to allow a MEB to convene and further evaluate a Marine's case.

d. Forwarding the case to CMC (MMSR-4) for review.

(1) The MTF must notify the Marine's command and must notify and forward a copy of the abbreviated limited duty form to CMC (MMSR-4) via naval message.

(2) Upon review, TLD may be approved, or the MEB may forward the case to the PEB for determination of fitness for continued military service.

e. Assigning the Marine to TLD. TLD is an assignment of limited duty for a specified period. Assignment to TLD can only be made after the completion of a MEB, and only if the prognosis is that the Marine can be RTFD within the specified period. The cumulative period of TLD will not exceed 16 months. The MCTFS duty limitation code for this status is "Q, LD MED BOARD" per paragraph 1206 of reference (i). The Marine's command is responsible for entering the code into MCTFS when TLD is authorized.

(1) Enlisted Assignment to TLD. A MEB recommending an initial period of TLD up to 8 months for enlisted Marines may be approved at the local MTF without approval from CMC (MMSR-4).

(2) Officer Assignment to TLD. A MEB recommending any period of TLD for Marine Officers must be submitted to CMC (MMSR-4) for review and approval.

(3) Medical Reevaluation. A MEB reassigned TLD period. The MEB at the local MTF must inform the command and CMC (MMSR-4) of the Marine's updated medical status prior to the completion of the TLD period.

(4) Assignment Extensions. Subsequent enlisted and officer assignment that extend the TLD status beyond 8 months must be submitted to CMC (MMSR-4) for review and approval.

f. Forwarding the case to a PEB for determination of fitness to continue military service. Chapter 8 of reference (h) and paragraph 2302 below further detail actions to be taken.

2302. PEB

1. The PEB acts on behalf of the SecNav to make determinations of fitness for continued military service, entitlement to benefits, disability ratings and disposition of the medical case. A PEB is required to convene for cases that exceed the maximum 16 months of limited duty or when it is not anticipated that a Marine will RTFD consequent to the initial diagnosis and prognosis. Reference appendix K of this Manual for further details on when a PEB is required.

2. PEB Findings

a. "Fit" for Continued Naval Service. To be found "Fit" to continue Naval service by the PEB, evidence must exist to demonstrate that the Marine is reasonably able to perform (without medical impairment) the duties of their billet, grade, or rank remaining on their period of service obligation to include Reserve obligation. Marines found "Fit" by the PEB are eligible for appropriate assignment. A finding of "Fit" does not preclude subsequent determinations of unsuitability for deployment, PFT participation, disqualification for special duties, TLD, or administrative action (including separation) resulting from such determination.

b. "Unfit" for Full Duty. To be found "Unfit" by the PEB to continue Naval service, evidence must exist to demonstrate that the Marine is unable to reasonably perform the duties of their billet, grade, or rank remaining on their period of service obligation to include Reserve obligation. The PEB will direct CMC (MMSR-4) to separate or retire Marines found "Unfit" to continue Naval service.

c. PLD. PLD is a continuation on active duty in a PLD status, which is approved after a Marine has been found "Unfit" by a PEB as a result of physical disability. PLD is assigned when a Marine is not expected to RTFD earlier than 16 months or it is projected that the condition cannot be corrected. A Marine can remain on active duty in a PLD status, provided the criteria (e.g., $\geq$ 18 and $\leq$ 20 years active duty) is met per reference (d). CMC (MMSR-4) may screen, review, and grant requests for PLD not meeting the mentioned criteria on a case-by-case basis.

(1) SRB/OQR Entry Requirements. When a permanent medical condition has been identified, appropriate entries will be placed in the Marine's health record and SRB/OQR. This will eliminate the necessity to repeatedly issue medical waivers. The health record entry will include MEB/PEB findings, including the date of the medical determination, diagnosis, prognosis, and restrictions imposed. The NAVMED 6100/5, Abbreviated Limited Duty Medical Board Report, will be filed on the document side of the SRB/OQR after completing the MCTFS entry requirements.

(2) MCTFS Entry. PLD is requested through the PEB and approved by CMC (MMSR-4). PLD requests will be authorized by the PEB based on the best interest of the Marine Corps and the health of the Marine. The MCTFS duty limitation code for this status is "S, RETN LD STAT" per paragraph 1206 of reference (i). Only CMC (MMSR-4) is authorized to enter or remove this code from MCTFS.

(3) FitRpt Requirement. Marines on limited duty for an entire semi-annual period will have "not medically qualified (NMED)" recorded in block 8b of section A and an explanation included in section I of the FitRpt per reference (d).

(4) A Marine assigned to PLD for a period greater than 6 months, must be reevaluated and have a complete MEB sent to the PEB no later than 4 months prior to PLD expiration date. A copy will be forwarded to the CMC (MMSR-4). This permanent condition will be reevaluated and updated at each service-related required physical evaluation (e.g., annual, reenlistment, etc.).

2303. BODY COMPOSITION STANDARDS AND LIGHT OR TLD. Marines assigned a light or TLD status or placed on medical hold for MEB/PEB screening will maintain body composition standards. Being screened for or placed on light or TLD does not absolve the Marine's responsibility to comply with Marine Corps standards. A Marine that fails to maintain body composition standards may be placed on a BCP, assigned to a modified RPCP (reflecting physical restrictions as determined by a MO), and given dietary adjustments to follow. Light or TLD Marines assigned to a BCP, who fail to successfully return to Marine Corps standards upon completion of the Program, are subject to administrative action to include a Page 11 counseling entry for unsatisfactory performance per paragraph 6105, and processed for administrative separation per reference (h).

2304. REELISTMENT/PROMOTION

1. Enlisted

a. Reenlistment. Enlisted Marines shall not be granted reenlistment while assigned to a PLD status unless the CMC grants a specific waiver. Instead, the Marine should be placed on medical hold, Convenience of the Government Medical (COFGM), until found "Fit" or "Unfit" by a PEB per paragraph 8607 of reference (h).

b. Promotion (TLD). Enlisted Marines assigned a TLD status do not lose their eligibility to be considered for promotion, due to medical conditions diagnosed by a MO that limits full duty status. Refer to paragraph 8608.8 of reference (k).

c. Promotion (PLD). Enlisted Marines who have been previously selected and are pending a fitness for duty determination by the Disability Evaluation System (DES) will be presented their certificates of appointment, and their promotions will be affected, if otherwise qualified. However, enlisted Marines assigned to a PLD status on the convening date of the appropriate selection board are not eligible for consideration for promotion.

2. Officers. If an officer is in a PLD status, delays of promotion can be initiated at the time an officer is selected by a selection board or prior to the time an officer's promotion is to be effected. Delay of an officer's promotion is warranted if the SecNav, CMC, or CO believes the officer is mentally, physically, morally, or professionally not qualified for promotion to the next higher grade. Refer to paragraph 8002.9 of reference (h) and paragraph 6003 of reference (m).

2305. PREGNANCY/POSTPARTUM. After medical determination of pregnancy, Marines will be waived from the PFT. Pregnant Marines will participate in a medically approved exercise program throughout the pregnancy and postpartum period, except when not recommended by a MO.

1. Resuming Physical Conditioning and Testing. Postpartum Marines will be required to take and pass a PFT no later than 6 months from the date that a MO has authorized her RTFD. This full duty authorization will normally coincide with the end of the 6-week (postpartum) convalescent leave period, per reference (l). Unless medically prohibited, physical conditioning should resume at this time. COs should be attentive to the fact that a Marine RTFD status will need to train at a moderate pace to return to the level of fitness she experienced prior to her pregnancy.

2. Early Termination/Miscarriage. A Marine whose pregnancy results in a miscarriage or has been terminated will be required to conform to body composition standards and pass a semi-annual PFT, in a time period determined by a MO. Just as with each pregnancy, recovery periods will vary depending on the individual's physical status at the time of termination/miscarriage. Time limits will not exceed those that are prescribed for a full term pregnancy.

3. Difficult Pregnancy or Birth/Cesarean Section/Still Birth. Although rare, these types of instances may result in a more difficult recovery. Therefore, in pregnancies where complications develop, MOs will determine on a case-by-case basis the time needed to return a Marine to full duty. Upon RTFD, the Marine will be required to conform to the body composition standards and pass the semi-annual PFT in the time period determined by a MO.

4. Lines of Communication. It is important that an open dialogue and coordination between the Marine, the MO, and the CO be maintained in order to determine any physical limitations and dietary concerns, the healing time required to return the Marine to full duty, and the Marine's overall progress.

5. Failure to RTFD. Marines not RTFD at the termination of the 6-week (postpartum) convalescent leave period, may be given an additional period of up to 30 days light duty, to allow additional recovery time. If upon termination of the 30 days light duty, the Marine is still not RTFD, then the Marine will be placed on medical hold and referred to a MEB for determination of fitness for continued military service. A Marine not RTFD status and as a result referred to a MEB will not be assigned to the BCP for failure to meet weight and body fat standards. Chapter 2, paragraph 2301 further details the MEB process and appendix L describes the Pregnancy and (inactive) BCP Assignment Process.

6. Promotion and Reenlistment

a. Pregnant Marines who did meet the reenlistment or promotion criteria prior to medical determination of pregnancy are eligible to reenlist and be promoted at any time during their pregnancy or (postpartum) convalescent period.

b. Pregnant Marines who did not meet reenlistment or promotion criteria prior to medical determination of pregnancy will not be eligible to reenlist or be promoted during the pregnancy period. Reenlistment and promotion eligibility will occur for these Marines only after they have been RTFD and have met all criteria.

2306. MEDICAL WAIVERS/EXEMPTIONS

1. Medical Waivers. A medical waiver (light or TLD) that excuses a Marine from participating in a PFT will not be considered a PFT failure. The MO will document on the medical evaluation the medical condition, the expiration date, physical exercise (based on limitations), and the PPFT events that the Marine can complete, if any, during the waiver period. Marines who are medically excused from performing a particular event of the PFT, as certified by a MO, are required to perform a PPFT. Reference paragraph 2202 of this Manual for PPFT requirements.

2. Injury During PFT. Marines injured during the execution of a semi-annual PFT (e.g., pulled muscle during the run) may be given credit for a PPFT score if two of the three events were completed prior to the injury. Or, after clearance from a MO, the Marine will be given an opportunity to run the PFT again. Only the CO and the Command PT Representative, after reviewing the facts surrounding the incident, can authorize that credit be given for a PPFT or if a Marine is given an opportunity to run the PFT at a later time. Marines completing only one event

prior to the injury will receive neither a PFT score nor PPFT credit. If evaluated and assigned a light duty status as a result of the injury, then the Marine will be required to retake the PFT 30-90 days after being RTFD status. Marines who are evaluated and not assigned a light duty status, will retake the PFT at the CO's discretion.

3. Illness. Marines encountering a rapid onset illness prior to the semi-annual PFT will obtain a light duty chit. Light duty chits must be submitted to the command prior to the scheduled PFT. The CO will decide when the Marine will be required to take the missed PFT.

2307. AGE 46 AND OVER RISK FACTOR SCREENING

>CH 1 1. Marines age 46 years and over will self-validate a "Risk Factor Screening" as a precautionary measure, 30-60 days prior to each semi-annual PFT. However, during IGMC/CGI team inspections, the inspectors may direct personnel age 46 and over to complete the Risk Factor Screening checklist if one is not on file. Marines will complete sections (A) and (B) of the Risk Factor Screening contained in appendix E, and submit the completed form to their CO for review. A medical screening by a MO, prior to PFT/PPFT participation, will only be required if the Marine:

a. Answers "yes" to any question in section (A) or (B).

b. Exceeds the body composition standards.

2. Marines requiring a follow-on medical screening must be medically cleared by a MO prior to participating in the PFT. Supplemental SF600 forms will be attached to the Screening if additional MO remarks are required. Independent Duty Corpsmen are not authorized to issue medical clearance to Marines who answered "yes" to any risk factor screening question. Marines who are not cleared by a MO to participate in the semi-annual PFT are restricted from participation during the current semi-annual period, either voluntarily or involuntarily. Such restriction only applies to the semi-annual period during which the MO determination was made.

3. Marines will participate in the PFT only if cleared by a MO after the referral, and no further medical evaluation is required. The signed and completed Risk Factor Screening (original) and SF600 documents will be filed in their health records, and a copy of the screening will be given to the CO. The Command PT Representative will annotate the Marine has completed the screening (been cleared for participation), on the PFT and Body Composition Worksheet.

4. Marines having a current physical examination or Navy Annual Health Evaluation (BUMEDINST 6120.26) will not be required to complete the semi-annual Risk Factor Screening or receive additional medical clearance to participate in the PFT if:

a. Examination or evaluation was conducted within the last 6 months from the date of the scheduled PFT.

b. There were no physical restrictions or limitations imposed by the MO.

c. Their state of health has not since changed.

CHAPTER 2

PHYSICAL FITNESS TEST (PFT)

SECTION 4: REMEDIAL PROCEDURES

2400. RPCP

1. The command directed RPCP is designed to improve a Marine's physical fitness and body composition, through command-structured physical conditioning, guidance, and supervision as contained in appendix C.

2. Assignment. Marines will be placed on a RPCP for any of the following:

a. Failing a semi-annual PFT.

b. Marines assigned to the BCP.

c. Marines required to take a semi-annual PFT but fail to do so ("required did not take, RDNT").

d. Marines who consistently demonstrate poor physical performance in MOS related tasks.

e. Female Marines assigned to the BCP prior to pregnancy and RTFD status will resume active participation in the RPCP. The level of program participation may require MO input, if physical limitations are identified.

3. Program Requirements. RPCP requirements:

a. Minimum of 30 days in the Program, regardless of change in ability.

b. Take and pass a PFT at completion of the initial 30-day period.

c. Marines assigned to the BCP will remain in the RPCP for the duration of the BCP assignment.

d. No formal administrative action is required for assignment to RPCP (e.g., no requirement for a SRB/OQR, Unit Diary, or MCTFS entry).

CHAPTER 3

BODY COMPOSITION PROGRAM (BCP)

	PARAGRAPH	PAGE
SCOPE	3000	3-3
OBJECTIVES	3001	3-3
RESPONSIBILITIES	3002	3-3
SECTION 1: STANDARDS		
BODY COMPOSITION STANDARDS	3100	3-5
BODY COMPOSITION EVALUATION PROCEDURES.	3101	3-5
PHYSICAL PERFORMANCE EVALUATION	3102	3-8
PREGNANCY/POSTPARTUM BODY COMPOSITION REQUIREMENTS	3103	3-9
SECTION 2: BCP PROCEDURES		
BODY COMPOSITION MEDICAL EVALUATION	3200	3-11
BCP ASSIGNMENT	3201	3-11
RESIDENTIAL OBESITY PROGRAM (ROP)	3202	3-18
TABLE		
3-1 ADMINISTRATIVE ACTIONS.		3-17

CHAPTER 3

BODY COMPOSITION PROGRAM (BCP)

3000. SCOPE. It is essential that every Marine maintain the established standards of health and physical fitness. With age, it is not uncommon for the average American to experience a decrease in physical activity, a change in metabolic rate, with unchanged caloric intake. As a result, body fat and weight may increase, having a negative impact on one's health and level of fitness. Medical authorities state that excess body fat can be associated with high blood pressure, high blood cholesterol, diabetes, cancer, cardiovascular disease, and similar health risks. In extreme climates and physically demanding environments, excess weight and body fat can also impede physical performance and stamina, which is of particular importance to the Marine and the mission.

3001. OBJECTIVES. The objectives of the Marine Corps' BCP are:

1. To establish healthy weight and body fat standards that ensure all Marines are physically capable to perform their duties.

2. To contribute to the health and well being of every Marine by continuously monitoring weight, body fat, diet/nutrition, and physical fitness conditioning.

3. To motivate all Marines to set the example by maintaining the established body composition standards.

4. To ensure those Marines who do not meet established standards are counseled accordingly, and given the opportunity and proper guidance to achieve the standards through the BCP and Remedial Physical Conditioning Program (RPCP).

3002. RESPONSIBILITIES

1. Individual Marine. It is every Marine's responsibility to maintain the Marine Corps' body composition standards, or if not in adherence with standards, to take appropriate action to return to them in a timely manner. Marines on light or limited duty, whose medical condition precludes them from participating in specific activities, will be expected to participate in conditioning alternatives and dietary adjustments, in order to maintain these standards.

2. CO. It is the CO's responsibility to ensure each Marine receives a semi-annual Body Composition Evaluation, and to closely monitor Marines assigned to the command's BCP and RPCP. COs will:

a. Continually monitor all Marines within the command to ensure adherence to established body composition standards.

b. Identify Marines who exceed the weight and body fat standards by conducting periodic Body Composition Evaluations.

c. Offer educational programs, events, and other motivational means to encourage Marines to achieve and maintain appropriate body composition standards. Such programs could include, but are not limited to: body composition seminars, health and fitness fairs, and individual consultations conducted by qualified dieticians, medical authorities, and Semper Fit personnel. Spouses should be encouraged to participate in these programs with the assigned Marine. Reference the Semper Fit website (www.usmc-mccs.org) for additional information and services.

d. Ensure the leadership understands that Marines may not have the education and experience needed to attain and maintain proper health and fitness, and that lifestyle changes may take time. However, if given the proper tools (e.g., education, supervision, and positive reinforcement), leaders can assist Marines in achieving and maintaining a healthy and fit lifestyle more quickly.

3. Medical Officer (MO). The MO's responsibility is to evaluate the Marine's body composition, overall health, diet, and physical ability to participate in the BCP and RPCP. The CO should receive the following from the MO:

a. Determination of whether the weight or body fat increase is due to an underlying cause or associated disease.

b. The Marine's current duty status (e.g., full, light, or limited duty).

c. The recommended dietary plan and weight and body fat reduction goals.

d. The physical conditioning restrictions or limitations imposed.

e. The date the Marine is expected to return to full duty (RTFD) status.

f. Any other pertinent information that pertains to the Marine's individual case.

CHAPTER 3

BODY COMPOSITION PROGRAM (BCP)

SECTION 1: STANDARDS

3100. BODY COMPOSITION STANDARDS. The Marine Corps' weight and body fat standards are health and performance based, and not based on appearance. Marines are considered not within these standards when their body weight and body fat exceed the maximum limits as contained in appendix I. If tested and the Marine's percent body fat exceeds the maximum limit (18% for males, 26% for females), then the CO will evaluate the Marine's physical performance (e.g., current semi-annual Physical Fitness Test (PFT)) using the Physical Performance Evaluation in paragraph 3102. If the Marine does not meet the Physical Performance Evaluation criteria, then the CO will have the Marine evaluated by a MO and assigned to a BCP - formerly known as the "Weight Control Program." Marines assigned to the BCP will receive assistance in reducing body weight and in particular body fat, in order to attain and maintain a more healthy physical fitness state. The BCP assignment process and endorsements are contained in paragraph 3201 and appendix D of this Manual.

3101. BODY COMPOSITION EVALUATION PROCEDURES

>CH 1 1. Body Composition Evaluation (BCE). The BCE is a semi-annual (January-June and July-December) requirement (annual for Reserves) designed to ensure Marines are within HT/WT or BF standards. However, commanders may direct BCE's for a unit or individuals as often as deemed necessary. BCE's may be conducted on the same day as the PFT. IGMC/CGI teams may also conduct unannounced BCE's in support of inspections.

a. Weigh-Ins. Regardless of duty status (full/light/limited duty), all Marines are required (at a minimum) to complete a semi-annual (annual for Reserves) Body Composition Evaluation as prescribed by the unit training schedule. The Command Physical Training (PT) Representative and his/her designate are the **only** unit personnel authorized to conduct Body Composition Evaluations, and will record the results on the PFT and Body Composition Worksheet in appendix F.

(1) Uniform. Uniform for the Evaluation is the USMC established green-on-green T-shirt and shorts, and socks. No other uniform or PT attire is permitted. Shoes will not be worn.

(2) Height Measurement. Height will be measured using a standard, non-stretching (cloth or fiberglass) tape measure that is secured to a wall and flush with the deck; weight scales that have a height measuring device will not be used. The Marine will stand with their back against the wall, heels flat on the deck, shoulders back, with arms to the side in a relaxed manner, and head straight forward. A ruler will be used to mark the Marine’s exact height.

(a) Measurement will be recorded to the nearest inch.

(b) If the height fraction is less than 1/2 inch, round down to the nearest inch. If height fraction is 1/2 inch or more, round up.

(3) Weight Measurement. Weight will be measured on a calibrated balance beam scale in the required PT uniform, without running shoes. A one-pound reduction will be granted for the PT uniform; however, no other weight reductions are authorized.

(a) Measurement will be recorded to the nearest pound.

(b) If the weight fraction is 1/2 pound or less, round down to the nearest pound. If more than 1/2 pound, round up.

(4) Refer to section I of appendix I to review the height and weight chart. If a Marine is not within height and weight standards, then the Marine will be measured for body fat immediately (same day). Body fat estimation is only authorized for Marines who exceed their maximum weight standard. The height and weight measurement conducted during this portion of the Body Composition Evaluation is what will be reported on the FitRpt.

b. Body Fat Estimation. Body fat will be estimated using the circumference-based method with one set of measurements for each gender (Males: height, neck circumference, and abdominal circumference at the navel. Females: height, neck circumference, waist circumference at the thinnest portion of the abdomen, and hips). No substitute methods of assessment are permitted. This DoD directed method of body fat estimation has been carefully evaluated for applicability to service members and represents the best approach that can be applied with minimal error (+/- 3 to 4%). This method is also valid because of the emphasis on abdominal circumference, the site of human body fat deposition which is most strongly associated with health risks, and which corresponds to other military goals including appropriate appearance and healthy exercise habits. Body fat measurements will be taken in the following manner:

(1) Marines will report for body fat estimation in the required PT uniform. The Command PT Representative and his/her designate will conduct the measurements.

(2) The tape measure used should be made of non-stretchable material, preferably fiberglass; cloth or steel tapes are unacceptable. The tape measure width should be 1/4 - 3/8 inch. The tape measure should be calibrated (i.e., compared with a yardstick or a metal ruler to ensure validity). This is done by aligning the fiberglass tape measure with the quarter inch markings on the ruler. The markings should match those on the ruler; if not, do not use that tape measure.

(3) All measurements will be taken on bare skin with the only exception being the female hip measurement.

(4) For use in the body fat estimation formula, Marines will have their height re-measured and recorded to the nearest 1/2-inch.

(a) If the measured height fraction is greater than an inch, round up to the nearest 1/2-inch.

(b) If the measured height fraction is greater than 1/2-inch, round up to the next whole inch.

(c) If the measured height fraction is at 1/2 inch, use that measurement in the body fat estimation equation.

(d) The height measured at the initial semi-annual (annual for Reserves) height/weight screen is what will be reported on the FitRpt, not the height used for the body fat estimation formula.

(5) Without causing indentation, the tape will be applied to the skin with sufficient tension to hold it in place as the measurement is being read.

(6) Circumference measurements will be taken three times to ensure accuracy. Each set of measurements will be completed sequentially to discourage assumption of repeated measurement readings. For males, complete one set of abdomen and neck measurements, **not**, for example, three abdominal circumferences followed by three neck circumferences. Continue the process by measuring the abdomen and neck in series until three sets of measurements are completed. For females, complete one set of waist, hip, and neck, measurements, **not**, for example, three waist measurements followed by three hip measurements. Continue the process by measuring the waist, hip, neck series three sets of measurements are completed. The lower of the three body fat measurements will be recorded on the PFT/Body Composition Worksheet in appendix F.

(7) Procedures for Body Fat Estimation for Males:

(a) Measure the neck circumference by placing the edge of the tape measure flush with the bottom of the larynx and perpendicular to the long axis of the neck. The Marine should look straight ahead during the measurement, with shoulders down (not hunched). For neck measurements in excess of the whole inch, round the neck measurement up to the nearest 1/2-inch and record (e.g., round up 16 1/4 inches to 16 1/2 inches).

(b) Measure abdominal circumference against the skin at the navel, level and parallel to the deck. Arms are at the sides. Take measurement at the end of the Marine's normal, relaxed exhalation. Round the abdominal measurement down to the nearest 1/2-inch and record (e.g., round down 34 3/4 to 34 1/2 inches).

(c) Determine percent body fat by subtracting the neck from the abdominal measurement and comparing this value against the height measurement, (e.g., abdominal - neck = circumference value). Refer to the chart "Percent Body Fat Estimation for Males" in section II of appendix I.

(8) Procedures for Body Fat Estimation for Females:

(a) Measure the neck circumference by placing the edge of the tape measure flush with the bottom of the larynx and perpendicular to the long axis of the neck. The Marine should look straight ahead during the measurement, with shoulders down (not hunched). To establish neck measurements in excess of the whole inch, round the neck measurement up to the nearest 1/2-inch and record (e.g., round up 13 3/8 inches to 13 1/2 inches).

(b) Measure the natural waist circumference against the skin at the point of minimal abdominal circumference, usually located about halfway between the navel and the lower end of the sternum (breastbone). When this site is not easily observed, take several measurements at probable sites and use the smallest value. Ensure the tape is level and parallel to the deck. Arms are at the sides. Take measurements at the end of a normal, relaxed exhalation. Round natural waist measurement down to the nearest 1/2-inch and record (e.g., round down 28 5/8 inches to 28 1/2 inches).

(c) Measure the hip circumference while facing the Marine's right side by placing the tape around the hips so that it passes over the greatest protrusion of the buttocks as viewed from the side. Ensure the tape is level and parallel to the deck. Apply sufficient tension on the tape to minimize the effect of clothing. Round the hip measurement down to the nearest 1/2-inch and record (e.g., round down 38 3/8 inches to 38 inches).

(d) Determine percent body fat by adding the waist and the hip measurements, subtracting the neck measurement, and comparing values against the Marine's height measurement (e.g., waist + hip - neck = circumference value). Refer to the chart "Percent Body Fat Estimation for Females" in section II of appendix I.

(e) Marines who meet the body fat standards, 18% for males and 26% for females, will be considered within the Marine Corps' body composition standards. Marines who exceed the body fat standards will be further evaluated on criteria set forth in paragraph 3102 below.

3102. PHYSICAL PERFORMANCE EVALUATION. There is an integral relationship between body composition and physical performance that warrants consideration when evaluating the "total" Marine. For instance, there are Marines who exceed weight and body fat standards, yet perform exceptionally well on the semi-annual PFT and combat fitness tests. These Marines demonstrate both exceptional cardiovascular endurance and upper body strength, with a body composition that may be identified as slightly exceeding the standards, but overall could be viewed as maintaining a reasonable body weight. In such cases, the Marine Corps' view on physical fitness as "essential to the day-to-day effectiveness and combat readiness" should be the overriding factor in these circumstances. The Physical Performance Evaluation not only takes body composition and physical fitness performance into consideration, it also takes into account the acknowledged variance in the body fat estimation technique ("taping" method). The Physical Performance Evaluation accommodates for the acknowledged margin of error in body fat estimation and testing, normally a variance of 3 to 4%. Reference (f) establishes a requirement to make such an accommodation in view of this variance. In essence, the Physical Performance Evaluation assesses the "total" Marine (e.g., body composition and recent semi-annual PFT performance as well as acknowledges the margin of error in body fat testing).

1. The Physical Performance Evaluation requires a Marine to meet the following criteria:

>CH 1 a. A 1st Class PFT taken a maximum of 90 days before or 30 days after the body composition evaluation.

b. Body fat estimation does not exceed the standard by more than 4%, (≤22% for males, ≤30% for females).

2. Marines exceeding 22 and 30% respectively or score a 2nd Class PFT or lower will not be eligible for the Physical Performance Evaluation under any circumstance. These Marines will automatically be processed for a BCP and RPCP assignment.

3. Partial Physical Fitness Tests (PPFT) will not be considered. Marines must be fit to take a PFT and score a 1st Class to be considered for the Physical Performance Evaluation.

4. No adverse administrative action will be taken for Marines meeting the Physical Performance Evaluation criteria. The Body Fat estimation percentage will still be reported on the FitRpt. The following is a directed comment in section I on the FitRpt meeting the Physical Performance Evaluation criteria:

"MRO meets the Physical Performance Evaluation criteria in MCO P6100.12 and is within standards."

5. Marines meeting the Physical Performance Evaluation criteria will only be exempt from administrative action for a particular 6-month (semi-annual) period (or annual for Reserves). Marines will be reevaluated at the next semi-annual PFT (annual for Reserves) and Body Composition Evaluation per paragraph 2100.

6. Physical Performance Evaluation examples:

a. Example 1: A male Marine, who exceeds his maximum allowable weight, is measured at 21% body fat, and scores a current 1st Class PFT has met the Physical Performance Evaluation criteria and would not be assigned to the BCP.

b. Example 2: A male Marine, who exceeds his maximum allowable weight, scores a current 1st Class PFT, but is measured at 24% body fat, does not meet the Physical Performance Evaluation body fat criteria, and therefore would be assigned to the BCP.

c. Example 3: A female Marine who exceeds her maximum allowable weight, is measured at 32% body fat, and scores a current 2nd Class PFT, does not meet the Physical Performance Evaluation body fat and PFT criteria, and therefore would be assigned to the BCP.

3103. PREGNANCY/POSTPARTUM BODY COMPOSITION REQUIREMENTS. Marines will be exempt from weight and body fat standards during pregnancy, but will maintain active participation in the Physical Conditioning Program (PCP) unless otherwise prescribed by her MO. Marines in a postpartum status will be required to meet weight and body fat standards no later than 6 months following a MO's authorization to RTFD. This authorization normally coincides with the completion of the 6-week (postpartum) convalescence leave period, per reference (l).

1. Miscarriage/Stillbirth. A Marine whose pregnancy terminates prematurely or that results in a stillbirth will meet weight and body fat standards within a time period determined by a MO. Due to the varying disposition of this medical situation, it is important that an open dialogue exists between the Marine, the MO and her CO to determine the time needed to recover and conform to body composition standards. The recovery period in this case, however, will not exceed the 6 months authorized for full-term pregnancies.

2. BCP Assignment. Marines who become pregnant while assigned to the BCP will remain on the Program in an "inactive" status.

a. The only Unit Diary entry required when a Marine becomes pregnant while assigned to a BCP is to report MCTFS Duty Limitation Code of "N, PREGNANCY" per paragraph 1206 of reference (i). An advisory message will be generated on the unit's Diary Feedback Report (DFR) as follows: "MARINE PLACED IN INACTIVE WT CNTL STATUS." All further advisories concerning the Marine's status in the BCP will be suppressed for the duration of the pregnancy and the medical related periods of limited duty immediately following the pregnancy.

b. Following the MO's medical evaluation, the Marine will be returned to either a full duty status, or reported to a non-medical related limited duty status. The Marine will then resume active participation in the BCP. All Marines will be assigned for a 6-month period, regardless of time spent on the Program prior to pregnancy. The unit will receive the following advisory message on their DFR: "MARINE RETURNED TO ACTIVE WT CNTL STATUS FOR SIX MONTHS."

3. Promotion Eligibility. Marines will not be eligible for promotion if they are in an "inactive" status for a BCP assignment per paragraphs 1204.3n and 6009.2e of reference (k). Only when the Marine returns to the weight and body fat standards and meets all other promotion requirements will she be promoted. Reference paragraph 2305 of this Manual for further details.

CHAPTER 3

BODY COMPOSITION PROGRAM (BCP)

SECTION 2: BCP PROCEDURES

3200. BODY COMPOSITION MEDICAL EVALUATION

1. Marines not within the body composition standards and do not meet the criteria for Physical Performance Evaluation will be referred to a MO to be evaluated. A medical evaluation and clearance, and all the Program endorsements contained in appendix D must be completed before a Marine can be assigned to a BCP. During the MO's evaluation, a determination must be made on whether or not a Marine's increase in weight and body fat is due to an underlying cause or associated disease, as follows:

>CH 1 a. Underlying Cause Defined. The following conditions may be considered as an underlying cause or associated disease which exempts a Marine (temporary basis) from meeting height/weight (HT/WT) or bodyfat (BF) standards and not being assigned to a BCP:

(1) Metabolic conditions, such as hypothyroidism, diagnosed by a medical officer (MO). The temporary exclusion will expire within 6 months of the metabolic condition being normalized through treatment or a limited duty/permanent disability board will be initiated.

(2) Medical conditions which require administration of medications that directly result in temporary weight gain.

>CH 1 b. Neuromusculoskeletal (NMS) Conditions. NMS conditions (e.g., sprained ankle, low back strain) that cause Marines to be placed in a not fit for full duty status may not be considered as an underlying cause or associated disease which exempts a Marine (temporary basis) from meeting HT/WT or BF standards. A NMS condition/not fit for full duty status determination also does not preclude assignment to a BCP. Assignment to a BCP for Marines in a not fit for full duty status secondary to a NMS condition will include referral from a MO or appropriately credentialed health care provider (ACHCP) to an appropriate agency for nutritional guidance and commencement of an alternative conditioning program.

>CH 1 c. Out of Standards Without Underlying Cause. When receiving the initial evaluation by the MO it is determined that no underlying cause or associated disease is the reason for the Marine's weight and body fat gain, then the Marine will continue forward with the BCP assignment process. The following medical recommendations will be made per an individual Marine's case:

(1) Recommend a realistic weight loss (e.g., pounds per week) and percent body fat reduction time line.

(2) Recommend caloric intake reduction and nutritional guidance (dietary measures).

(3) Recommend alternative physical conditioning exercises based on physical limitations (if necessary) for Marines on light or limited duty.

>CH 1 d. Out of Standards With an Underlying Cause. When receiving their initial evaluation by the MO, it is determined that an underlying or associated disease is the reason for the Marine's weight and body fat gain, then it must be thoroughly

annotated in the Marine's Health Record and on the Program's First Endorsement. The following courses of action may be taken per individual Marine's case:

(1) Treatment to alleviate the medical condition and return of the Marine to the command.

(2) Hospitalization for obesity treatment at an Alcohol Rehabilitation Center (ARC). Criteria for ARC Residential Obesity Program (ROP) are contained in paragraph 3202 in this Manual.

(3) Diagnosed as untreatable or permanent in nature, case forwarded to a PEB for determination of fitness to continue military service.

(4) RTFD, assignment to BCP and RPCP.

3201. BCP ASSIGNMENT. The following actions will be taken by the command when a Marine is being screened for a BCP assignment. Marines assigned to the BCP will be required to participate in the Program for a minimum of 6 months.

1. Preliminary Medical Evaluation. The CO will complete the BCP Evaluation Letter (appendix D, enclosure 1) requesting a preliminary medical evaluation. This Letter and the First Endorsement (appendix D, enclosure 2) will be forwarded to the MO at the local Medical Treatment Facility (MTF). The Marine will report directly to the MTF to be medically evaluated by the MO.

2. First Endorsement. The MO will complete the First Endorsement (appendix D, enclosure 2) and return the Letter to the CO.

a. The MO will attach the following information to the Endorsement:

(1) Determination of whether the weight and body fat increase is due to an underlying cause or associated disease;

(2) Recommended weight and body fat reduction goals;

(3) Recommended dietary measures;

(4) Physical limitations or restrictions imposed; and

(5) Medical clearance to participate in the Command's BCP and RPCP.

b. If the Marine is diagnosed with an underlying cause or associated disease, then the MO may recommend that the Marine not be assigned to a BCP until the Marine's condition is treated, or may recommend that the Marine be processed for medical separation for untreatable conditions. This diagnosis must be thoroughly annotated in the Marine's Health Record and on the Program's First Endorsement. Reference paragraph 3200 in this Manual for further details regarding "underlying cause and/or associated disease."

c. The signature block on the Endorsement will clearly state the name, grade, Branch of Service, medical title, and organization (as appropriate) of the MO who diagnosed the cause of the Marine's deficiency.

3. Second Endorsement. Upon reviewing the MO's diagnosis and recommendations, the CO will complete the Second Endorsement (appendix D, enclosure 3). The CO will counsel the Marine on the deficiencies, corrective actions required, and information regarding the BCP and RPCP, and will have a 6105 counseling entry recorded on the Marine's SRB/OQR's Page 11 per reference (h).

4. Third Endorsement. Upon completion of the counseling, the Marine will sign and date the Third Endorsement (appendix D, enclosure 4) acknowledging the deficiencies, corrective actions required, and requirements for the 6-month BCP and RPCP assignments.

5. Admin Section Responsibilities. The CO will direct the Admin Section to:

a. On receipt of the Third Endorsement (appendix D, enclosure 4):

(1) Enter the BCP assignment on the Unit Diary and record the Unit Diary number and date on the bottom of the Third Endorsement. Forward this Endorsement to the Command PT Representative to be maintained in the Marine's BCP file.

NOTE: A BCP assignment will not be run on the Unit Diary until the Marine has been evaluated and cleared for participation by a MO, the appropriate Program endorsements have been completed, and the Marine has been properly counseled by the CO.

(2) Use the following code/statement to report assignment to BCP in MCTFS: TTC 337 000 TO WT CNTL.

b. When directed, record 6105 counseling entry on SRB/OQR Page 11.

c. When directed, submit a request for Marine Corps Institute (MCI) Course 3316, "Basic Nutrition."

d. Notify the Command PT Representative when the Admin Section has received a Diary Feedback Report (DFR) that identifies a Marine's BCP assignment has expired.

e. When directed, make an entry on the Unit Diary extending the Marine's initial (first) BCP assignment and enter the following code/statement per reference (e) into MCTFS: TTC 337 001 EXTEND WT CNTL. Record the Unit Diary number and date on the Sixth Endorsement (appendix D, enclosure 7) and forward it to the Command PT Representative to be maintained in the Marine's BCP file.

f. When directed, remove the Marine from the BCP, making an entry on the Unit Diary and the following code/statement per enclosure (e) into MCTFS: TTC 339 000 FR WT CNTL. Record the Unit Diary number and date on the Sixth Endorsement (appendix D, enclosure 7) and return it to the Command PT Representative to be maintained in the Marine's BCP file.

6. Command PT Representative Responsibilities. The CO will direct the Command PT Representative to:

a. On receipt of the Third Endorsement (appendix D, enclosure 4) from the Admin Section, create a BCP file for the Marine. The BCP file will contain all Marine's Program Endorsements, medical documentation, counseling, body composition testing results, training log, etc.

b. Immediately initiate the 6-month RPCP for a Marine medically qualified. A Marine in a light or limited duty status will be required to participate in an alternative RPCP, based on the physical restrictions and limitations imposed by the MO.

c. Conduct diet and nutrition classes for a Marine or set up appointments with the fitness and medical professionals at the local Semper Fit facilities or MTFs.

d. Order (via Admin Section) MCI Course 3316, "Basic Nutrition," for the Marine (whether Officer or Enlisted), and direct the Marine to complete the Course prior to BCP assignment expiration.

e. Take weight and body fat measurements once every 2 weeks (once a month SMCR), and document results in the Marine's BCP file.

f. Document in the Marine's BCP file the counseling conducted for those Marines who have displayed neither weight loss nor body fat reduction, and if it is a result of the Marine not adhering to the regimen prescribed by the MO and the command.

(1) A Marine not making satisfactory progress during the BCP assignment will receive a 6105 counseling entry on the SRB/OQR's Page 11 for failure to make satisfactory progress. This Page 11 entry will be given for unsatisfactory performance per paragraph 6206 in MCO P1900.16F, and is required for administrative separation.

(2) A Marine who makes progress, but fails to meet the weight and body fat standards at completion of a BCP assignment will receive a 6105 counseling entry on the SRB/OQR's page 11. This Page 11 entry will be given for failing to meet body composition standards per paragraph 6215 in MCO P1900.16F, and is required for administrative separation.

g. Maintain and update accordingly, the BCP file, (e.g., counseling, body composition test results, and RPCP training log entries).

>CH 1 h. Sgts and above who are assigned to the BCP will be administered an adverse FitRpt per reference (d). Cpls and below will receive commensurate PRO/CON marks per reference (g).

7. Fourth Endorsement. After a Marine has participated in the BCP for 4 months, the CO will complete the Fourth Endorsement (appendix D, enclosure 5) and will forward this endorsement and the Fifth Endorsement (appendix D, enclosure 6) to the MO at the local MTF. The Marine will report to the MTF to be reevaluated by the MO. (If at any time the Marine is failing to lose weight or body fat, the Marine should be evaluated earlier, in order to be reexamined for a possible underlying or associated disease.)

8. Fifth Endorsement. The MO will reevaluate the Marine, complete the Fifth Endorsement (appendix D, enclosure 6), and forward the endorsement to the CO with one of the following diagnoses/recommendations:

a. Underlying Cause or Associated Disease. The Marine has been diagnosed as having an underlying or associated disease, and the Marine will:

(1) Be placed in an "inactive status" (duty limit "N") for the BCP assignment until the Marine's condition is treated; or

(2) Be recommended for (medical) separation for untreatable conditions.

b. Reference paragraph 3200 for further guidance on cases involving an underlying cause or associated disease.

c. No Underlying Cause or Associated Disease and Making Progress. The Marine has been diagnosed as not having an underlying cause or associated disease and is making satisfactory progress towards weight and body fat reduction goals. Recommend that participation in the BCP (and RPCP) continue until assignment expires.

d. No Underlying Cause or Associated Disease - Not Making Progress. The Marine has been diagnosed as not having an underlying cause or associated disease and is not making satisfactory progress towards weight and body fat reduction goals. Recommend that participation in the BCP (and RPCP) continue until assignment expires.

(1) A Marine may not receive an extension if lack of satisfactory progress is a result of not adhering to the regimen prescribed by the MO and the command.

(2) A 6105 counseling entry on the Marine's SRB/OQR's Page 11 will be administered at this time (fourth month) for failure to make satisfactory progress. See paragraph 6206 in MCO P1900.16F.

e. Returned to Weight or Body Fat Standards. The Marine has met weight and/or body fat reductions goals and is within body composition standards. Recommended that participation in the BCP (and RPCP) continue until assignment expires.

9. Expiration of Program Assignment. The Admin Section will receive a DFR when the BCP assignment has expired. One of the following actions will be taken:

a. A Marine has met and maintained the weight and body fat standards and will be officially removed from the BCP (and RPCP); see paragraph 13 in this section for Program removal instructions.

b. A Marine has failed to return to the weight and body fat standards, but has made satisfactory progress. Refer to paragraph 11 below for further guidance on Program (first assignment) extensions.

c. A Marine has failed to return to the weight and body fat standards and has not made satisfactory progress. CO will administer a 6105 counseling entry on the Marine's SRB/OQR's Page 11, per paragraph 6206 or 6215 in MCO P1900.16F respectively, and will process the Marine for administrative separation.

10. Extensions to First Assignment. A Marine making satisfactory progress and reasonably close to meeting their weight and body fat standards at the end of the 6-month BCP assignment may be granted a Program extension.

a. CO can grant a one-time extension of up to 6 months to first assignments only. An extension to a first assignment does not constitute a second assignment.

b. A Marine granted an extension will continue participation in the BCP and RPCP until the extension period expires. A Marine granted an extension will be counseled pursuant to paragraph 6105 of reference (h) with standard language in subparagraph (1) below and have the following code statement entered into MCTFS: TTC 337 001 EXTEND WT CNTL. Counseling pursuant to paragraph 6105 of reference (h) for Marines not making significant progress while on their first assignment or the extension to their first assignment to the BCP will occur after the initial 60-day probation period and, at a minimum, at the fourth month of the BCP assignment when the Marine is reevaluated by the MO and diagnosed as not having an underlying cause or disease. Subparagraph (2) provides the recommended standard language for this counseling.

(1) "Counseled this date concerning your extension on the Marine Corps BCP. You have made reasonable progress but have failed to reach your required body composition while on your first assignment to this Program. Therefore, you have been granted a one-time extension of 6 months to meet the body composition standards set forth in MCO P6100.12. You are advised that failure to take corrective action and meet your body composition standards or weight/ body fat reduction goals, as prescribed by the medical officer, while on this 6 month extension will result in administrative separation from the USMC for either weight control failure or unsatisfactory performance per paragraphs 6206 and 6215 of MCO P1900.16E. Specific recommendations for corrective action are: Meet your weight/body fat reduction goals by losing the prescribed __ lbs and __ percent body fat per month for 6 months and actively participate in your diet and exercise program. Assistance is available through your chain of command. I was advised that within 5 working days after acknowledgment of this entry a written rebuttal could be submitted and that such a rebuttal will be filed on the document side of my SRB/OQR. I choose (to)(not to) make such a statement."

(2) "Counseled this date concerning your unsatisfactory performance while assigned to the Marine Corps BCP. Due to insufficient effort, you have not met your weight/body fat reduction goals as prescribed by the medical officer. You are advised that failure to take corrective action and reach your required body composition will result in administrative separation from the USMC for either weight control failure or unsatisfactory performance per paragraphs 6206 and 6215 of MCO P1900.16E. Specific recommendations for corrective action are: Meet your weight/body fat reduction goal and actively participate in your diet and exercise

program. Assistance is available through your chain of command. I am advised that within 5 working days after acknowledgment of this entry, a written rebuttal could be submitted and that such a rebuttal will be filed on the document side of my SRB/OQR. I choose (to)(not to) make such a statement."

>CH 1 c. Unlike the mandatory 6 month time period associated with the initial assignment to a BCP, Marines on an extension to a first assignment will be removed from the BCP when they return to HT/WT or BF standards; see paragraph 13 in this section for program removal instructions. A Marine not meeting standards at the end of the extension will be processed for administrative separation for failure to meet body composition standards per paragraph 6215 in MCO P1900.16F. No additional extensions can be granted.

>CH 1 11. Second Assignment. A Marine who has been assigned to and officially removed from the BCP at any time in their career, and fails to meet the weight and body fat standards a second time, may receive a one-time second assignment to the BCP (and RPCP). On a second assignment, Marines will be given 6 months with no extensions to conform to standards and will be counseled pursuant to paragraph 6105 of MCO P1900.16F with the recommended standard language. Unlike the mandatory 6 month time period associated with an initial assignment to a BCP, Marines on a second assignment to a BCP will be removed from the BCP when they return to HT/WT or BF standards. A Marine will be assigned to the Program as per paragraphs 2 through 10 of this section. A Marine successfully meeting the weight and body fat standards at the completion of the second assignment will be officially removed from the BCP (and RPCP) as stated in paragraph 13 in this Section. A Marine not meeting the standards when the second assignment expires will receive a 6105 counseling entry on the their SRB/OQR's Page 11 per paragraph 6206 or 6215 in MCO P1900.16F respectively, and will be processed for administrative separation. Extensions to a second assignment are not authorized.

> "Counseled this date concerning your reassignment to the Marine Corps BCP. Specifically, you failed to properly maintain your body composition standards as required by MCO P6100.12 for a second time. Your actions are unsatisfactory and not in accordance with the high state of readiness required by the USMC. You are advised that this subsequent assignment is for a 6-month period. Specific recommendations for corrective action are: Meet your weight/body fat reduction goals by losing the prescribed __ lbs and __ percent body fat per month and actively participate in your diet and exercise program. Assistance is available through your chain of command. You are advised that failure to take corrective action and reach the Marine Corps body composition standard will result in administrative separation from the USMC for either weight control failure or unsatisfactory performance per paragraphs 6206 and 6215 of MCO P1900.16E. Additionally, you are advised that MCO P6100.12 precludes assignment to the BPC for a third time in a Marine's career and that any subsequent failure to meet prescribed weight/body fat standards during your career will result in administrative separation. I am advised that within 5 working days after acknowledgment of this entry, a written rebuttal could be submitted and that such a rebuttal will be filed on the document side of my SRB/OQR. I choose (to)(not to) make such a statement."

12. Follow On Assignments. A Marine, assigned to the BCP on two separate occasions (e.g., first and second assignments), that fail to meet weight and body fat standards a third time will automatically be processed for administrative separation per paragraph 6215 of MCO P1900.16F. In such cases, the Marine will still be required to participate in the BCP and RPCP until discharged; no formal assignment is required. A prior counseling entry pursuant to paragraph 6105 of MCO P1900.16F is required to separate a Marine for either weight control failure or unsatisfactory performance per paragraphs 6206 and 6215 of MCO P1900.16F.

13. Program Removal. A Marine will not be removed from the BCP (and RPCP) assignment prior to the Program expiration date. A Marine will continue to participate in the BCP and RPCP as prescribed, regardless of progress, in order to ensure healthy weight and body fat loss techniques were employed and the weight and body fat loss is long term. A Marine successfully completing the assignment or extension will be administratively removed from the BCP (and RPCP).

a. Sixth Endorsement (appendix D, enclosure 7). The CO will complete the Sixth Endorsement, counsel the Marine on having successfully completed the BCP assignment, and have the Marine sign the endorsement. The Seventh Endorsement (appendix D, enclosure 8) only applies to Marines completing an extension to a first assignment.

b. The CO will direct the Admin Section to officially remove the Marine from the BCP, recording the following code/statement per reference (e) into MCTFS: TTC 339 000 FR WT CNTL. The Admin Section will record the Unit Diary number and date on the bottom of the endorsement and forward it to the Command PT Representative to be maintained in the Marine's BCP file.

c. Copies of completed BCP files will be kept on record and maintained by the Command PT Representative for 1 year.

14. Summary of Administrative Actions. Table 3-1 lists the administrative actions required for Marines who fail to meet the weight and body fat standards.

ADMINISTRATIVE ACTION	NUMBER OF ASSIGNMENTS	
	BCP 1st	BCP 2nd
Mandatory RPCP	YES	YES
Page 11 SRB/OQR Mandatory Entry	YES*	YES*
Adverse FitRpt or Substandard PRO/CON Marks	YES	YES
Eligible for Promotion	NO	NO
Eligible for PCS Transfer	YES	NO
Process for Administrative Separation	YES	YES
May Be Denied Re-enlistment	YES	YES
May Be Denied Special Schools	YES	YES
May Be Denied Recommendation for Promotion	YES	YES

Table 3-1.--Administrative Actions.

*A 6105 counseling entry will be recorded on the SRB/OQR's Page 11 the day the Marine is counseled on deficiencies, corrective actions to be taken, and assignment to the BCP. Additional Page 11 entries, per paragraph 6206 of MCO P1900.16F will only be required for a Marine who have not made satisfactory progress during the BCP assignment.

NOTE: A Marine who meets the Physical Performance Evaluation criteria will not be assigned to the BCP and will not receive an adverse FitRpt.

15. BCP and End of Active Service (EAS). CO discretion will be exercised when there are individual cases involving Marines who fail to meet weight and body fat standards just prior to their EAS.

a. Imminent EAS does not preclude a Marine who exceeds weight and body fat standards from being evaluated and processed for a BCP and RPCP assignment. A CO may evaluate and assign a Marine to participate in the BCP and RPCP up to the Marine's EAS date.

b. For a Marine who fails to meet weight and body fat standards upon completion of their 6-month BCP assignment or extension, and EAS is imminent, CO discretion will be exercised to decide whether or not measures to process the Marine for administrative separation is worth the administrative burden. CO may opt to extend the Marine's BCP and RPCP assignment up to the Marine's EAS date. A 6105 counseling entry on the SRB/OQR's Page 11 will be recorded, identifying the BCP failure, for Reserve component purposes.

16. Relocation Orders for Marines on the BCP

a. A Marine can be transferred, receive PCS or PCA orders during a first assignment, or (first assignment) extension to a BCP.

b. A Marine will not be transferred, but placed in an administrative hold status if he/she is currently participating in a second assignment to the BCP.

c. The Admin Section and Command PT Representative are required to ensure all Unit Diary entries, MO's Body Composition Evaluation(s), and BCP documents, are up to date and included in the Marine's BCP file, prior to being forwarded to the receiving command.

d. The command is required to forward by registered/receipt mail, the original BCP file to the command to which the Marine is being transferred, and keep a copy of the file on record for 1 year. This BCP file should be forwarded within 7 days of the Marine's official checkout date.

3202. Residential Obesity Program (ROP)

1. Marines that have been evaluated by a MO and determined to have an extreme weight and body fat increase due to an underlying or associated disease, will be recommended to participate in the ROP. The ROP is available to all Marines regardless of pay grade. A Marine will be sent to the ROP at the local ARC based on the following criteria:

a. A Marine has not previously participated in a ROP at a residential ARC during Marine Corps career.

b. A Marine has no medical history of having an eating disorder, (e.g., anorexia, bulimia, or other unresolved eating disorder).

c. A Marine must have a strong potential for continued military service supported by CO or related letters of recommendation.

d. A Marine must have at least 12 months time in service remaining on current military contract after the completion of the ROP for aftercare. [Waivers to remaining time in service criteria may be requested on a case-by-case basis from CMC (MMSR-4).]

2. Post-Residential Obesity Program (Rehabilitation). Upon completion of the ROP, the Marine will be assigned to the unit's RPCP for a minimum of 6 months. Weight and body fat reduction goals will vary depending upon the Marine's circumstances. Results must be monitored by the Command PT Representative and the MO assigned to the case. Entry in the ROP does not affect the BCP assignment process.

3. Rehabilitation Failure. In cases of rehabilitative failure, a Marine will be processed for administrative separation per paragraph 6206.1 of MCO P1900.16F. All rehabilitation failures must be reported to CMC (MMSR-4) via Naval message. Rehabilitation failure is based on the following circumstances:

a. Failure to demonstrate significant and consistent progress towards meeting Marine Corps weight and body fat standards during the 6-month rehabilitation period.

b. Continued increase in weight and/or body fat during the ROP or the aftercare period.

APPENDIX A

GLOSSARY

SECTION 1: MCO SHORT TITLES

MARCORMAN	THE MARINE CORPS MANUAL
MCO P5600.31G	MARINE CORPS PUBLICATIONS AND PRINTING REGULATIONS
MCO P1610.7E	PERFORMANCE EVALUATION SYSTEM (SHORT TITLE: PES)
MCO P1080.40C	MARINE CORPS TOTAL FORCE SYSTEM PERSONNEL REPORTING INSTRUCTIONS MANUAL (SHORT TITLE: MCTFSPRIM)
MCO P1070.12K	MARINE CORPS INDIVIDUAL RECORDS ADMINISTRATION MANUAL (SHORT TITLE: IRAM)
MCO P1900.16F	MARINE CORPS SEPARATION AND RETIREMENT MANUAL (SHORT TITLE: MARCORSEPMAN)
MCO P1080.20M	MARINE CORPS TOTAL FORCE SYSTEM CODES MANUAL (SHORT TITLE: MCTFSCODESMAN)
MCO P1040.31H	ENLISTED CAREER PLANNING AND RETENTION MANUAL
MCO P1400.31B	MARINE CORPS PROMOTION MANUAL, VOLUME 1, OFFICER PROMOTIONS (SHORT TITLE: MARCORPROMMAN, VOL 1, OFFPROM)
MCO P1400.32C	MARINE CORPS PROMOTION MANUAL, VOLUME 2, ENLISTED PROMOTIONS (SHORT TITLE: MARCORPROMMAN, VOL 2, ENLPROM)
MCO 5000.12D	MARINE CORPS POLICY ON PREGNANCY AND PARENTHOOD

SECTION 2: SHORT TITLES & ACRONYMS

APPROPRIATELY CREDENTIALED HEALTH CARE PROVIDER	ACHCP
ADMINISTRATIVE SEPARATION	ADSEP
ALCOHOL REHABILITATION CENTER	ARC
BODY COMPOSITION PROGRAM	BCP
CONVENIENCE OF THE GOVERNMENT	CofG
DISABILITY EVALUATION SYSTEM	DES
DIARY FEEDBACK REPORT	DFR
MARINE CORPS TOTAL FORCE SYSTEM	MCTFS
MEDICAL EVALUATION BOARD	MEB
MEDICAL OFFICER	MO
MILITARY TREATMENT FACILITY	MTF
NOT MEDICALLY QUALIFIED	NMED
PHYSICAL CONDITIONING PROGRAM	PCP
PHYSICAL EVALUATION BOARD	PEB
PERMANENT LIMITED DUTY	PLD
PARTIAL PHYSICAL FITNESS TEST	PPFT
PHYSICAL READINESS TRAINING	PRT
REQUIRED BUT DID NOT TAKE	RDNT
RESIDENTIAL OBESITY PROGRAM	ROP
REMEDIAL PHYSICAL CONDITIONING PROGRAM	RPCP
RETURN TO FULL DUTY	RTFD
TEMPORARY LIMITED DUTY	TLD

SECTION 3: DEFINITIONS

Appropriately Credentialed Health Care Provider (ACHCP). An individual who has specialized training or education in a health-related field. This may include administration, direct provision of patient care, or ancillary services. Such a professional is licensed, certified, or registered by a Government agency to provide specific health services in the field as an independent. An ACHCP, referred to throughout MCO P6100.12 as Medical Officer (MO), is a physician, physician's assistant, or nurse practitioner; Independent Corpsmen or regular Corpsmen are not considered ACHCPs.

Body Composition Program (BCP). Formerly known as the "Weight Control Program." Marines assigned to the BCP will receive assistance in reducing weight/body fat, in order to resume to a more healthy physical fitness state. The BCP assignment process and endorsements are contained in appendix D; the unit's administrative actions are contained in paragraph 3201 of MCO P6100.12.

Body Fat (BF). The amount of fat tissue a person has in ratio to lean body mass (or muscles, bones, organs, nervous tissue, and skin).

Diary Feedback Report (DFR). A report received in the Admin Section requiring action (e.g., DFR to notify the command that SNM's BCP assignment period has expired).

Light Duty. This is the status a MO assigns to a Marine when it is determined that a medical condition exists that interferes with the performance of duty, but the Marine is expected to return to full duty (RTFD). A MO may recommend up to 30 days light duty. When the light duty period expires, SNM will be medically reevaluated. If SNM is not RTFD status, then the Marine will be placed on medical hold and a Medical Evaluation Board (MEB) must be initiated and completed within the following 30 days.

Limited Duty. This is the status a MO assigns to a Marine when it is determined that a medical condition exists and will require an extended period of time before being RTFD. See Temporary Limited Duty and Permanent Limited Duty.

MEB. A MEB is a group of MOs at a local Military Treatment Facility (MTF) convened for the purpose of identifying the physical and mental qualifications of referred Marines to RTFD. The MEB is charged with determining whether a Marine's limitation(s) preclude their RTFD within a reasonable period of time or at all. A MEB will evaluate and report on the diagnosis, prognosis, possible RTFD, plan for further treatment, and make medical recommendation(s) for disposition.

MO. An ACHCP; see ACHCP.

Partial Physical Fitness Test (PPFT). Marines who are medically excused from performing a particular event of the PFT, as certified by a MO, are required to perform a PPFT. Marines will be required to obtain at least the minimum passing performance for the two events they are medically qualified, in order to pass the PPFT.

Permanent Limited Duty (PLD). PLD is a continuation on active duty in a permanent limited duty status, which is approved after a Marine has been determined "Unfit" by a PEB as a result of physical disability. PLD is assigned when a Marine is not expected to RTFD earlier than 16 months or it is projected that the condition cannot be corrected. A Marine can remain on active duty with a PLD status, provided criteria is met per MCO P1900.16F. PLD is requested through the PEB and approved by CMC (MMSR-4). PLD requests will be authorized by the PEB based on the best interest of the Marine Corps and the health of the Marine.

Physical Evaluation Board (PEB). The PEB acts on behalf of the SecNav to make determinations of fitness for continued military service, entitlement to benefits, disability ratings, and disposition of the medical case. A PEB is required to convene for cases that exceed the maximum 16 months of limited duty or when a Marine's initial diagnosis and prognosis do not anticipate a return to RFTD.

Physical Fitness Test (PFT). An instrument that measures the level of physical fitness of all Marines. It is a measurement of general fitness vice combat readiness and MOS capability. The standard PFT consists of three events that measure cardiovascular endurance, muscular strength and endurance, and mobility. Male Marines will perform dead-hang pullups, abdominal crunches, and a 3.0 mile run. Female Marines will complete the flexed arm hang, abdominal crunches, and a 3.0 mile run. These events are designed to test the strength and stamina of the upper body (shoulder girth), midsection, and lower body, as well as the efficiency of the cardiovascular system.

Physical Readiness Training (PRT). This training is comprised of general physical conditioning exercises that will develop and maintain strength, endurance, and the physical skills necessary to sustain individual Marines during combat.

Remedial Physical Conditioning Program (RPCP). Remedial physical conditioning is a process by which Marines are conditioned in a progressive manner to meet prescribed standards. The Program affords Marines the opportunity to rebuild themselves after a weight/body fat increase, injury/illness, pregnancy, or a period that lacked a structured fitness program.

Residential Obesity Program (ROP). Marines that have been evaluated by a MO and determined to have an extreme weight/body fat increase due to an underlying cause or associated disease, will be recommended to participate in the ROP.

Temporary Limited Duty (TLD). TLD is an assignment of limited duty for a specific period. Assignment to TLD can only be made after the completion of a MEB, and only if the prognosis is that the Marine can be RTFD within the specified period. The cumulative period of TLD will not exceed 16 months.

APPENDIX B

DAILY 16 PROGRAM

A. General

1. Purpose. The Daily 16 Program is a comprehensive series of warmup, conditioning, and cool-down exercises; it replaces the former Daily 7. The Daily 16 warmup exercises facilitate gradual distribution of blood flow to the muscles, preparing both the cardiovascular and musculoskeletal systems for the exercise session, by effectively targeting both the upper and lower body. The increased blood flow to the muscles produces a warming effect, increasing the elasticity of the muscles and connective tissues, which is believed to reduce injury risks. The Daily 16 conditioning exercises include traditional calisthenics that are safe in providing a total body workout, which can vary in duration, degree of difficulty, and level of intensity. The Daily 16 cool-down exercises (the same exercises used in the warmup) allow the body to gradually return to the pre-exercise state.

2. Scope. There are three different ways the Daily 16 Program can be used during physical training:

a. Daily 16 Warmup (D16WU). A series of warmup and dynamic stretching exercises that should be conducted prior to the main activity (e.g., formation run, obstacle course, circuit course, etc.), of every physical training session.

b. Daily 16 Workout (D16WO). A series of conditioning exercises that can be used as augmentation to another conditioning activity (circuit course, Physical Fitness Test (PFT), etc.), or can be used as a conditioning activity in of itself by simply increasing the number of repetitions or by slowing down the execution of the repetition.

c. Daily 16 Cool-Down (D16CD). A series of cool-down exercises that should be used as the final activity of a physical training session.

3. Daily 16 Warmup and Dynamic Stretching Descriptions

a. Warmup Exercises

(1) Heel to Toe Rocking. Starting position is standing with feet together and hands on hips. Rock back onto the heels, pause, then rock forward onto the toes and pause. Repeat 10 to 15 repetitions.

1

2

(2) Partial Squats. Starting position is standing with feet shoulder width apart, arms at the sides (1). Keeping the heels on the deck, partially squat until hands are near mid-calf. The knees should only bend to about 60 degrees, just short of a sitting position (2). Return to the starting position (3). Repeat 10 to 15 repetitions.

1/3

2

(3) Butt Kicks. Starting position is standing with feet shoulder width apart, hands on hips. Shift weight onto the right foot and quickly bend the left knee five times, bringing the left heel toward the buttocks. Switch legs and repeat. Repeat the whole cycle two or three times, until a total of 15 to 20 repetitions are done on each leg.

(4) Double Time in Place. Starting position is standing with arms at the sides. Slowly begin to run in place and gradually increase speed. While double-timing, conduct "punches to the front (1)" (throw easy punches to the front of the body). Switch to "punches to the sky (2)" (throw easy punches straight up to the sky). Finish with "arm circles (3)", (begin small then transition to large arm circles; repeat in other direction).

1 2 3

(5) Neck Flexion and Extension. Starting position is standing with feet shoulder width apart, hands on hips. Flex the neck forward bringing the chin toward the chest, pause. Extend the head back and pause. Repeat for 5 to 10 repetitions.

(6) Neck Lateral Flexion. Starting position is standing with feet shoulder width apart, hands on hips. Tilt the head to the left side, bringing the left ear toward the left shoulder, pause. Switch sides and repeat. Repeat for 5 to 10 repetitions.

(7) Trunk Flexion and Extension. Starting position is standing with feet shoulder width apart, hands on hips (1). Extend the trunk backward bringing a slight hyperextension of the trunk and pause (2). Repeat for 5 to 10 repetitions.

1

2

(8) Trunk Lateral Flexion. Starting position is standing with feet shoulder width apart, hands on hips. Bend the trunk to the left side (1), pause, then switch to the right side (2) and pause. Repeat for 5 to 10 repetitions. Proceed to the Stretching Exercises.

1

2

b. Dynamic Stretching Exercises

(1) Upper Back Stretch. Starting position is standing with feet shoulder width apart. Extend the arms and clasp the hands in front of the chest. Push the arms forward rounding the shoulders and upper back. Hold the position for 10 seconds and breath naturally. The stretch should be felt over the upper back.

(2) Chest Stretch. Starting position is standing with feet shoulder width apart. Clasp hands together behind the lower back, palms up. Pull the arms up toward the head. Hold the position for 10 seconds and breath naturally. The stretch should be felt in the front of the chest and shoulders.

(3) Modified Hurdler Stretch. Starting position is in the sitting position. Extend the left leg out while tucking the right leg in front of the hips with the knee pointing outward. Bend the torso forward toward the left knee. The stretching should be felt over the back of the left thigh. Hold the position for 10 seconds and breath naturally. Switch sides and repeat.

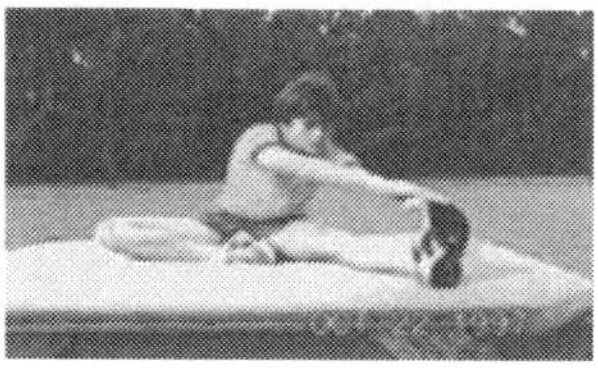

(4) Hip and Back Stretch. Starting position is in the sitting position. Extend the right leg straight out and cross the left leg over the right leg by bending the left knee and placing the left foot on the deck next to the right knee. Turn the upper torso to the left pushing the left knee to the right with the right elbow. Hold the position for 10 seconds and breath naturally. The stretch should be felt over the lower back and left hip. Switch sides and repeat.

(5) Groin Stretch. Starting position is in the sitting position with both knees bent and the bottoms of the feet together. Grasp the feet and gently push the knees with the elbows toward the deck. Hold the position for 10 seconds and breath naturally. The stretch should be felt over the inside of both thighs.

(6) Calf Stretch. Starting position is standing with arms at the sides. Place the left foot approximately 2 feet forward and slightly bend the right knee. Lean forward toward the left foot pointing the left toes up to the sky. Hold the position for 10 seconds and breath naturally. Grabbing the left foot and gently pulling it towards you can increase the level of intensity. The stretch should be felt over the left calf. Switch sides and repeat.

(7) Iliotibial Band (ITB) Stretch. Starting position is standing with arms at the sides. Place the left foot behind and a few inches to the right of the right foot. Bring the left arm over the head. Place your body weight on the left leg and bend at the waist to the right. Hold the position for 10 seconds and breath naturally. The stretch should be felt over the left hip. Switch sides and repeat.

(8) Hamstring Stretch. Starting position is lying down with the back flat against the deck. Bring the left knee toward the chest grasping the left leg just below the knee. Gently straighten the left knee and hold for the count. The right leg should remain on the deck. Hold the position for 10 seconds and breath naturally. The stretch should be felt on the back of the left thigh. Switch sides and repeat.

(9) Neck Stretch. Starting position is standing with feet shoulder width apart. Move both arms behind the back and grasp the left wrist with the right hand. Tilt the head to the right and pull the left arm to the right. Hold the position for 10 seconds and breath naturally. The stretch should be felt over the left shoulder and left side of the neck. Switch sides and repeat.

(10) Triceps Stretch. Starting position is standing, arms at the sides. Bend the left elbow and bring the left arm up and back placing the left hand between the shoulder blades. Gently pull the left elbow with the right hand behind the head. Hold the position for 10 seconds and breath naturally. The stretch should be felt over the back of the upper arm. Switch sides and repeat.

(11) Quadriceps Stretch. Starting position is lying down on the left side. Bend the left hip and knee to 90 degrees. Grasp the right ankle with the right hand and pull the right knee straight back. Do not hyperextend the lower back. Hold the position for 10 seconds and breath naturally. The stretch should be felt over the front of the right thigh. Switch sides and repeat.

(12) Lying Down ITB Stretch. Starting position is lying down on the deck. Bring the left leg with the knee straight across the body (1). Hold the position for 10 seconds and breath naturally. The stretch should be felt over the left hip. Switch sides and repeat (2).

1

2

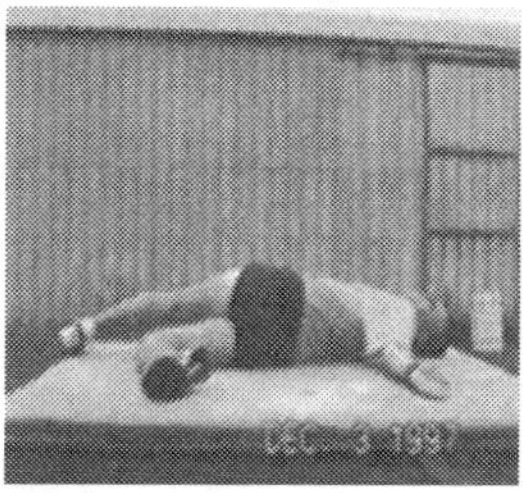

(13) Posterior Shoulder Stretch. Starting position is standing with feet shoulder width apart, arms at the sides. Bend the left elbow and bring the left arm across the chest. Give a gentle pull with the right hand. Hold the position for 10 seconds and breath naturally. The stretch should be felt over the posterior left shoulder. Switch sides and repeat.

(14) Hip Flexor Stretch. Starting position is standing, hands on hips. Step the left foot forward 3 to 4 feet. Place the right knee on the deck. Gently move the left knee forward. Hold the position for 10 seconds and breath naturally. The stretch should be felt over the front of the right thigh and hip. Switch sides and repeat.

(15) Single-Leg Lower Back Stretch. Starting position is lying with the back flat against the deck. Bring the right knee toward the chest grasping the right knee. Gently pull the knee tight into the chest. The left leg should remain on the deck. Hold the position for 10 seconds and breath naturally. The stretch should be felt along the lower back to the right buttock. Switch sides and repeat.

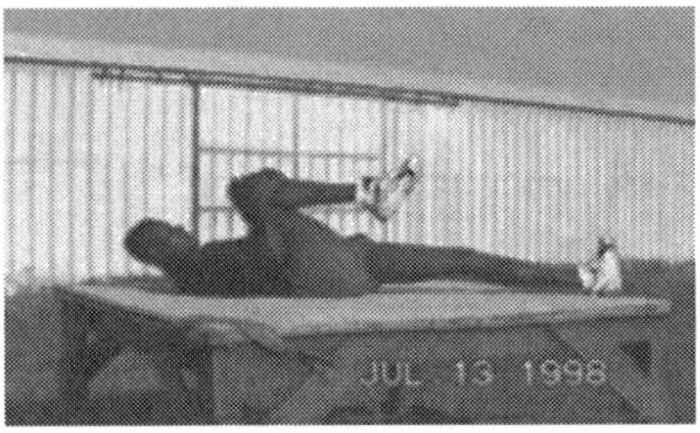

(16) Prone Abdominal Stretch. Starting position is lying on the stomach with the hands placed near the shoulders as if in the down position of a pushup. Slowly raise the upper body, keeping the waist on the deck. Hold the position for 10 seconds and breath naturally. The stretch should be felt over the abdomen.

4. Daily 16 Conditioning Exercise Descriptions

a. Pushups. Starting position is lying on your stomach with hands shoulder width apart, toes on the deck, and elbows, back, and knees straight. On the first and third count, lower the chest to the deck; bend the elbows to at least 90 degrees (1/3). On the second and fourth count, extend the arms back to the starting position (2/4). This exercise conditions the chest, primarily in the anterior shoulder region, and secondarily, the triceps.

1/3

2/4

b. Crunches. Starting position is lying on the back with the hips bent to 90 degrees and the knees bent, feet off the deck. Bend the elbows to 90 degrees and fold across the chest or rib cage. On the first and third count raise the upper torso off the deck touching the thighs with the forearms (1/3). On the second and fourth count, return to the starting position (2/4). The arms must remain in constant contact with the chest/rib cage when executing the crunch. This should be done in a slow and controlled manner. This exercise conditions the abdominal muscles.

1/3

2/4

c. Dirty Dogs. Starting position is on the hands and knees. On the first and third count raise the left leg to the side, while keeping the knee bent (1/3). On the second and fourth count return the leg to the starting position (2/4). Switch sides and repeat. This exercise conditions the hip abductors.

1/3

2/4

d. Wide Pushups. Starting position is lying on the stomach with hands beyond shoulder width apart, toes on the ground, and elbows, back, and knees straight. On the first and third count, lower the chest to the deck, bending the elbows to at least 90 degrees (1/3). On the second and fourth count, extend the arms back to the starting position (2/4). This exercise conditions the chest, primarily in the anterior shoulder region, and secondarily, the triceps. With the wider hand position, the chest muscles increase their workload.

1/3

2/4

e. Dive Bomber Pushups. Starting position is lying on the stomach with hands and toes on the deck, and elbows and knees straight. The hands will be slightly beyond shoulder width apart. The hips will be raised up and the shoulders will be behind the hands. On the first count, lower the chest down and forward to the deck, the shoulders will be even with the hands (1). On the second count, continue forward extending the elbows where now the shoulders are in front of the hands (2). On the third count, reverse the direction lowering the chest down and back to the deck, the shoulders will be even with hands (3). On the fourth count, continue back and up to the starting position (4). This exercise is done in a smooth, continuous motion.

This exercise conditions the chest and anterior shoulder primarily, and secondarily, the triceps through a greater ranger of motion.

1

2

3

4

f. <u>Elbow to Knee Crunches</u>. Starting position is lying on the back with the right foot flat on the deck and, the left foot crossed over the right knee, and the arms crossed over the chest. On the first and third count, raise the upper torso off the deck rotating to the left touching the right elbow to the left thigh (1/3). On the second and fourth count, return to the starting position (2/4). This exercise should be done in a slow and controlled manner. Switch sides and repeat. This exercise conditions the abdominal muscles with more emphasis on the oblique.

1/3

2/4

g. <u>Side Crunches</u>. Starting position is lying on the left side with the left arm across the chest and right arm along the side of the body. On the first and third count, raise the upper torso and feet off the deck sliding the right hand down the thigh (1/3). On the second and fourth count, return to the starting position (2/4). Switch sides and repeat. This exercise conditions the abdominal muscles with emphasis on the internal oblique and external oblique.

1/3

2/4

h. <u>Prone Flutter Kicks</u>. Starting position is lying on the stomach. On the first count raise the left leg off the deck while the right leg remains on the deck (1). On the second count, lower the left leg to the starting position (2). On the third count, raise the right leg off the deck while the left leg remains on the deck (3). On the fourth count, lower the right leg to the starting position (4). This exercise conditions the muscles that extend the hip and back.

1

2

3

4

i. Back Extension. Starting position is lying on the stomach with hands behind the head. On the first and third count, raise the upper torso and legs off the deck (1/3). On the second and fourth count, lower the upper torso and legs to the starting position (2/4). This exercise conditions the muscles that extend the back.

1/3

2/4

j. Donkey Kicks. Starting position is on the hands and knees. On the first and third count, kick the left leg back and up straightening the knee (1/3). On the second and fourth count, bend the knee and hip bringing the left knee into the chest (2/4). The back should not hyperextend during this exercise. Switch sides and repeat. This exercise conditions the muscles that extend the hip.

1/3

2/4

k. Hip Adduction. Staring position is lying on the left side with the right leg bent setting the right foot in front of the left knee. On the first and third count, raise the straight left leg off the deck squeezing the thighs together (1/3). On the second and fourth count, lower the left leg to the starting position (2/4). The left toes should be pointing straight-forward not to the sky. Switch side and repeat. This exercise conditions the muscles on the inner thigh.

1/3

2/4

1. Side Leg Raises. Starting position is lying on the left side with the left knee bent, the hip vertical and the toes on the right foot pointing forward not to the sky. On the first and third count, raise the right leg approximately 18 inches leading with the heel (1/3). The toes will still point forward, not to the sky. On the second and fourth count, lower the right leg to the starting position (2/4). Switch sides and repeat. This exercise conditions the muscles on the side of the hip and thigh.

1/3

2/4

m. Steam Engines. Starting position is standing with the feet shoulder width apart and hands behind the head. On the first count, touch the right elbow to the left knee by bending and raising the left knee and twisting and bending the upper torso to the left (1). On the second count, return to the starting position (2). On the third count, touch the left elbow to the right knee by bending and raising the right knee and twisting and bending the upper torso to the right (3). On the fourth count, return to the starting position (4). This exercise should be done in a slow and controlled cadence. This exercise conditions the abdominal muscles, primarily the oblique, and secondarily, the hip and thigh.

1

2

3

4

n. Lunges. Staring position is standing with the feet shoulder width apart and hands on the hips. On the first count, touch the right knee to the deck by stepping forward with the left foot and bending both knees (1). On the second count, return to the starting position (2). On the third count, touch the left knee to the deck by stepping forward with the right foot and bending both knees (3). On the fourth count, return to the starting position (4). Do not bend the forward knee more than 90 degrees. This exercise should be done in a slow and controlled cadence. This exercise conditions the muscles that extend the hip and knee of the forward leg.

1

2

3

4

o. Side Straddle Hops. Starting position is standing with the feet together and arms at the sides. On the first and third count, jump and land with both feet just beyond shoulder width apart while bringing the hands together overhead (1/3). On the second and fourth count, jump back to the starting position (2/4). This exercise is done in a smooth, continuous motion. This exercise conditions the upper arm, hip, and thigh.

1/3

2/4

5. Daily 16 Warmup/Cool-Down, Stretching and Exercise Cards

STRETCHING CARD A

- Chest Stretch
- Triceps Stretch
- Posterior Shoulder Stretch
- Iliotibial Band (ITB) Stretch
- Modified Hurdler Stretch
- Hip and Back Stretch
- Quadriceps Stretch
- Lower Back Stretch
- Abdominal Stretch

EXERCISE CARD 1

- Wide Pushups
- Donkey Kicks
- Crunches
- Dive Bomber Pushups
- Dirty Dogs
- Side Crunches
- Back Extensions
- Lunges
- Side Straddle Hops

STRETCHING CARD B

- Upper Back Stretch
- Chest Stretch
- Shoulder & Neck Stretch
- Triceps Stretch
- Posterior Shoulder Stretch
- Lying Down (ITB) Stretch
- Modified Hurdler Stretch
- Groin Stretch
- Calf Stretch

EXERCISE CARD 2

- Pushups
- Crunches
- Side Leg Raises
- Diamond Pushups
- Elbow-to-Knee Crunches
- Prone Flutter Kicks
- Hip Abduction
- Lunges
- Steam Engines

STRETCHING CARD C

- Active Hamstring Stretch
- Groin Stretch
- Hip Flexor Stretch
- Lower Back Stretch
- ITB Stretch
- Calf Stretch
- Neck Stretch
- Upper Back Stretch
- Chest Stretch

WARM-UP/COOL-DOWN CARD

- Toe-Heel Rocking
- Partial Squats
- Butt Kicks
- Trunk Bends
- Neck Bends
- Run in Place
- Punch to the Front
- Punch to the Sky
- Arm Circles
- *Stretch Card (10 Seconds)
- *Exercise Cards

APPENDIX C

REMEDIAL PHYSICAL CONDITIONING PROGRAM (RPCP)

1. Purpose. Remedial physical conditioning is a process by which Marines are conditioned in a progressive manner to meet prescribed physical fitness and body composition standards. The RPCP goal is to provide challenging conditioning sessions, using the spectrum of aerobic conditioning, resistance conditioning, and other related exercises that will afford the Marines an opportunity to rebuild themselves after a weight or body fat increase, injury/illness, pregnancy, or a period that lacked a structured fitness program.

2. Assignment to RPCP. Participation in the RPCP is a mandatory requirement for Marines who fail the Physical Fitness Test (PFT) or who are assigned to the Body Composition Program (BCP). Marines will also be assigned to the RPCP when they consistently display problems in MOS-specific tasks, (e.g., conditioning marches, etc.) or struggle to meet the minimum standard for an event during a semi-annual PFT. Performance on unit formation runs will not be used as RPCP entry/exit criteria. Marines recovering from injuries or prolonged illness may be assigned to the RPCP if a formal remedial program would be beneficial. In the later case, COs must take into account any followup medical care (therapy) requirement and tailor the RPCP to accommodate those physical limitations or restrictions imposed by the Medical Officer (MO). Deficiencies, which can be improved or corrected by a command-directed remedial conditioning program, fall into several categories:

a. Lack of strength in one or more body parts;

b. Lack of overall cardiovascular or muscular endurance;

c. Lack of adherence to body composition standards;

d. Lack of mobility (posture, balance, stability, agility, coordination, power, speed and flexibility);

e. Lack of motivation and dedication; and/or

f. Lack of education and experience in health and fitness.

3. Organization of the RPCP

a. Leadership. The intent of the RPCP is not punitive in nature, but designed to encourage improvement in overall fitness and body composition. Unit leaders and Command Physical Training (PT) Representatives should assist Marines in clearly identifying their deficiencies, developing specific strategies to improve, and tracking progress. The RPCP should focus on realistic goal setting, healthy life-styling strategies, progressive PT, dietary measures, and testing and evaluation. The RPCP should be challenging, employ a cross-training philosophy, and ensure safety measures are taken to avoid over-training and potential overuse injuries.

b. Use of Resources. This Appendix will provide a few examples of remedial conditioning programs units can employ or use as a guide for their own program designs. Command PT Representatives can also access the resources available at Semper Fit Fitness Centers, which are staffed by qualified professionals and offer a variety of health and fitness services.

c. Organization. Remedial conditioning should be conducted in small groups, (e.g., no larger than 10-15 Marines per group), thereby allowing for more focused instruction, oversight, and reinforcement. It is recommended that the RPCP initially be organized by grouping Marines by area of deficiency and level thereof. Prescribed exercise loads should be commensurate with their abilities, and whenever possible, linking a strong Marine with a weaker Marine. An organization as such will support performance and improvement.

(1) As the RPCP and individual fitness and performance levels progress, Marines should be reorganized to concentrate on remaining deficiencies, or combined to conduct specific PT events, such as conditioning marches or team sports. For example, a group that exhibits a deficiency in upper body strength would execute a program that includes pullups, rope climbing, rifle PT, log drills, and weight training. A second group may exhibit deficiencies in cardiovascular respiratory endurance, and therefore would execute a program that includes individual and ability group running, grass drills, circuit training, and fartlek running.

d. Measurement. Continual measurement of a Marine's deficient area(s) is the only way to gauge progress. While testing should not take the place of training sessions geared towards improving performance, periodic testing (e.g., every 2 weeks) is an important step of any remedial conditioning program. Testing, PFT, and body composition evaluations for example, should commence at the end of the first week of a RPCP.

4. RPCP

a. Circuit Training and Exercise Drills. Circuit training is described as a specific training routine in which a variety of fitness discrepancies are addressed. It can accommodate a large number of Marines with little supervision, provide variety and challenge with its progressive programming, and allow participants to progress at their own rate. The goal of circuit training is to develop strength and endurance through a systematic and progressive conditioning program that involves stations where specific exercises are performed. These exercises are performed vigorously for a short period of time before moving to a follow-on station. The vigorous activity in short periods of time provide a near maximal quality training session, assuring progressive overload in an organized manner. The exercise selected for each station and the arrangement of the stations is determined by the objective of the circuit.

(1) Free Circuit. Free (portable) circuits use equipment that can be readily transported or set up for use regardless of the training area and environment. A simplified course can be executed on a basketball court, a field, or on the deck of a ship. In a free circuit, there is no set time for staying at each station and no signal to move from one station to the next. Marines work at their own pace, doing a fixed number of repetitions (rep) at each station. Progress is measured by the time needed to complete a circuit. However, in remedial training, the quality and number of repetitions should be monitored.

(2) Fixed Circuit. Fixed circuits require equipment that is normally in place and can be used on a regular basis. In a fixed circuit, a specific length of time is set for each station. The time is monitored with a stopwatch, with Marines rotating through the stations on command. To increase the intensity or difficulty of a fixed circuit:

(a) maintain the time for completion, but increase the number of reps;

(b) increase the time per station along with the number of reps; and,

(c) increase the number of times a circuit is negotiated.

b. Number of Stations/Circuit Training Reps. The goal of the circuit, as well as the time and equipment available will determine the number of stations. A circuit geared for the objective of developing upper or lower body strength may need as little as six to eight stations, while a cardiovascular strength and endurance circuit may have as many as 20 stations. The number of Marines assigned to negotiate a single circuit station is determined by the total number of participants, the number of stations, and the level of supervision available. To achieve the desired training effect, the circuit may have to be repeated several times. For example, a 10-station circuit requiring Marines to exercise for 30 seconds at each station and 15 seconds to move between stations may result in Marines completing the entire evolution three times. The entire training session, designed to take between 45-60 minutes, includes a warmup and cool-down period (e.g., Daily 16). As the conditioning level increases, consideration can be given to either increase exercise periods from 30 to 45 and 60 second intervals, or the addition of a fourth rotation through the circuit. Stations should be arranged in a sequence that allows for adequate recovery time between high/low intensity exercises or by alternating different muscles groups.

(1) Cardiovascular Endurance Circuit #1. This circuit is designed to improve the cardio-respiratory system by exercising hard at each of the eight exercise stations, with a steady recovery time (slow jog) between stations. Recommended course length is approximately 1/4 to 1/2 mile in length.

Station #1:	Ab Crunch	20 reps
Station #2:	Bend and Thrust	20 reps
Station #3:	Sprint	1 rep
Station #4:	Inclined Pushups	20 reps
Station #5:	Star Jumps	20 reps
Station #6:	Dorsal Raise	20 reps
Station #7:	Knee Lifts	20 reps
Station #8:	Dips	20 reps

(2) Cardiovascular Endurance Circuit #2. Recommended course length is 1/2 to 1.0-mile, with 14 events/stations:

Station #1:	Stationary Run	30 seconds
Station #2:	Pushup	30 seconds
Station #3:	Side Straddle Hop	30 seconds
Station #4:	Crunches	30 seconds
Station #5:	Ski Jumps	30 seconds
Station #6:	Flutter Kicks	30 seconds
Station #7:	Bend and Reach (slow reps)	30 seconds
Station #8:	Wide Pushups	30 seconds
Station #9:	Bicycle	30 seconds
Station #10:	Knee Bender	30 seconds
Station #11:	Steam Engines (Daily 16)	30 seconds
Station #12:	Twisting Situp	30 seconds
Station #13:	Lunges (Daily 16)	30 seconds
Station #14:	All-Fours Run	30 seconds

(3) Upper Body Strength Circuit #2. This circuit is designed to improve upper body strength and endurance, and is normally conducted following a cardiovascular training session. Proper form and full range of motion are important. Eight stations are set up over a course length of approximately 500 yards.

Station #1:	Crunches	20 reps
Station #2:	Pullups	3 sets @ max effort
Station #3:	Dorsal Raises	20 reps
Station #4:	Tricep Dips	40 reps
Station #5:	Situps	2 sets of 5 reps
Station #6:	Rope Heaves	20 reps
Station #7:	Pushups	20 reps
Station #8:	Sprint	1 rep

(4) Upper Body/Abdominal Strength Circuit #3. This circuit is designed to encompass a cardiovascular/anaerobic (sprint) portion that incorporates eight exercise stations with no set course length required. This circuit is conducted without a set time signal, e.g., Marines move immediately to a station after each exercise is completed. Recommended one to two complete course rotations, with a possible time decrease to 20 seconds (vice 30) on the second running.

Station #1:	Elevated Pushup	30 seconds
Station #2:	Twisting Situp	30 seconds
Station #3:	Parallel Dips	30 seconds
Station #4:	Ab Crunch (feet unsecured)	30 seconds
Station #5:	Wide Pushup	30 seconds
Station #6:	Flutter Kicks	30 seconds
Station #7:	Diamond Pushup	30 seconds
Station #8:	Bicycle	30 seconds

(5) Lower Body/Upper Body Strength (Weight Room) Circuit #4. This circuit is designed to follow a cardiovascular session. There are 13 exercise stations and the circuit should be conducted in one to two complete rotations. Each exercise should encompass slow and controlled movements, striving for muscle failure within 8-12 repetitions. A preliminary session, for Marines to become familiar with weight room layout and starting weight limits and execution is recommended.

Station #1:	Leg Press	8-12 reps
Station #2:	Leg Raise	8-12 reps
Station #3:	Leg Extension	8-12 reps
Station #4:	Leg Curl	8-12 reps
Station #5:	Heel Raise	8-12 reps
Station #6:	Bench Press	8-12 reps
Station #7:	Seated Row	8-12 reps
Station #8:	Military Press	8-12 reps
Station #9:	Lat Pulldown	8-12 reps
Station #10:	Shrugs	8-12 reps
Station #11:	Triceps Extension	8-12 reps
Station #12:	Biceps Curl	8-12 reps
Station #13:	Incline Situp	8-12 reps

c. Fartlek Training. Fartlek training is designed to develop strength, stamina, and endurance by combining running (cardiovascular) and calisthenics (strength and endurance). Fartleks consist of various running exercises, which strive to overload the cardiovascular system by allowing only partial recovery between each intense training segments; this sequence is repeated for the duration of the training session. Fartleks are normally conducted over a minimum 3-mile course, preferably with varying terrain and intermittent exercise stations to create the work overload.

(1) Initial/Baseline Fartlek:

- Warmup (Daily 16 warmup routine)
- Warmup run at easy pace for 5-10 minutes
- Run fast, steady pace for 3/4-1 3/4 miles
- Walk/Jog at moderate pace for 5 minutes (recovery)
- Run easily - sprinting 15-20 meters occasionally
- Run full speed uphill for 175-200 meters
- Cool-down by running easily for 1/2-1 mile
- Cool-down (Daily 16 cool-down routine)

(2) Cardiovascular/Muscular Endurance Fartlek: Recommended course length is 3.0-3.5 miles. Location and spacing of exercise stations is normally based on terrain. After a unit orientation run, the Fartlek course should be run as an individual effort.

Station #1:	Mountain Climbers	20 reps
Station #2:	Bends and Thrust	20 reps
Station #3:	Ab Crunches (feet unsecured)	20 reps
Station #4:	Sprint	175-200 meters
Station #5:	Situp w/Twist	20 reps
Station #6:	Pullups or Pushups	10-20 reps
Station #7:	Dorsal Raises	20 reps
Station #8:	Star Jumps	20 reps
Station #9:	Squat Thrusts	20 reps

5. RPCP Schedules

a. Upper Body Strength (Weekly) Circuit. A 3 day per week routine is recommended to improve overall upper body strength deficiencies. Conduct tests every 2 weeks and increase weights and reps as necessary.

Monday: Weight Training

Pullups	2 sets to fatigue
Lat Pulldown*	2 sets of 15 reps
Seated Row*	2 sets of 15 reps
Biceps Curl (EZ-Curl Bar)*	2 sets of 15 reps
Alternating Bicep Curls (Dumbbell)	2 sets of 15 reps

* Strive for momentary muscular fatigue at the end of each set. Allow 45-60 seconds rest between sets and 2-3 minutes rest between exercises.

Tuesday: Cross Training - Cardiovascular/Abdominal Training

Wednesday: Pullup/Pushup/Crunch Pyramid **

Set #1:	1 Pullups/2 Pushups
Set #2:	2 Pullups/4 Pushups
Set #3:	3 Pullups/6 Pushups/25 Crunches
Set #4:	4 Pullups/8 Pushups
Set #5:	5 Pullups/10 Pushups/25 Crunches
Set #6:	4 Pullups/8 Pushups
Set #7:	3 Pullups/6 Pushups/25 Crunches
Set #8:	2 Pullups/4 Pushups
Set #9:	1 Pullups/2 Pushups

** Decrease/Increase reps in pyramid based on the Marine's ability and progression of strength, the minimal rest between sets, and spotter assistance necessary to complete required reps.

Thursday: Rest

Friday: Repeat Monday training session

b. <u>Strength and Endurance Training Schedule (20 weeks)</u>. This training schedule's primary fitness element is running, but also strives to combine aspects of strength, endurance, and flexibility. The running portion of the program provides the base of cardiovascular endurance necessary to augment all other physical skills. In addition to the progressive overload running program, strength training and flexibility exercises (e.g., Daily 16), addressing lower and upper body resistance training must be incorporated. Weight training can augment or be substituted for the Daily 16 exercises. Each training period will include a warmup prior to exercise and a cool-down session to follow.

WEEK 1,2	Endurance	Strength
Monday	Run 40 min @ 10 min/mile pace	Lower Body Resistance Daily 16 or Weight Training (1 set)
Tuesday	None	Upper Body Resistance - Daily 16 or Weight Training (1 set)
Wednesday	Run 30 min @ 10 min/mile pace	Lower Body - (1 set)
Thursday	None	Upper Body - (1 set)
Friday	Run 30 min @ 10 min/mile pace	Lower Body - (1 set)

WEEK 3,4	Endurance	Strength
Monday	Run 44 min @ 10 min/mile	Lower Body - (1 set)
Tuesday	None	Upper Body - (1 set)
Wednesday	Run 33 min @ 10 min/mile	Lower Body - (1 set)
Thursday	None	Upper Body - (1 set)
Friday	Run 33 min @ 10 min/mile	Lower Body - (1 set)

WEEK 5,6	Endurance	Strength
Monday	Run 48 min @ 10 min/mile	Lower Body - (2 sets)
Tuesday	None	Upper Body - (2 sets)
Wednesday	Run 36 min @ 10 min/mile	Lower Body - (2 sets)
Thursday	None	Upper Body - (2 sets)
Friday	Run 36 min @ 10 min/mile	Lower Body - (2 sets)

WEEK 7,8	Endurance	Strength
Monday	Run 53 min @ 10 min/mile	Lower Body - (2 sets)
Tuesday	None	Upper Body - (2 sets)
Wednesday	Run 40 min @ 10 min/mile	Lower Body - (2 sets)
Thursday	None	Upper Body - (2 sets)
Friday	Run 40 min @ 10 min/mile	Lower Body - (2 sets)

WEEK 9	Endurance	Strength
Monday	Run 53 min @ 9 min/mile	Upper Body - (3 sets)
Tuesday	None	Lower Body - (3 sets)
Wednesday	Run 40 min @ 10 min/mile	Upper Body - (2 sets)
Thursday	None	Lower Body - (3 sets)
Friday	Run 40 min @ 10 min/mile	Lower Body - (2 sets) Power - 2 reps

WEEK 10	Endurance	Strength
Monday	Run 53 min @ 8.5 min/mile	Upper Body - (2 sets)
Tuesday	Intervals: 4x 400m sprints in 2 min	Lower Body - (3 sets)
Wednesday	Run 40 min @ 8.5 min/mile	Upper Body - (2 sets)
Thursday	None	Lower Body - (3 sets) Power - 3 reps
Friday	Run 40 min @ 8.5 min/mile	Upper Body - (2 sets)

WEEK 11	Endurance	Strength
Monday	Run 53 min @ 8.5 min/mile	Upper Body - (3 sets)
Tuesday	Intervals: 4x 400 m sprints in 2 min	Lower Body - (2 sets)
Wednesday	Run 40 min @ 8.5 min/mile	Upper Body - (3 sets)
Thursday	None	Lower Body - (2 sets) Power - 4 reps
Friday	Run 40 min @ 8.5 min/mile	Upper Body - (3 sets)

WEEK 12	Endurance	Strength
Monday	Run 58 min @ 8.5 min/mile	None
Tuesday	Intervals: 5x 400 m sprints in 2 min (4 min rest)	Upper/Lower Body - (2 Sets)
Wednesday	Run 40 min @ 8.5 min/mile	None
Thursday	None	Upper/Lower Body - (3 sets) Power - 4 reps
Friday	Run 30 min @ 7.5 min/mile	None

WEEK 13	Endurance	Strength
Monday	Run 58 min @ 8.5 min/mile	None
Tuesday	Intervals: 6x 400 m sprints in 2 min (4 min rest)	Upper/Lower Body - (3 sets)
Wednesday	Run 40 min @ 8.0 min/mile	None
Thursday	None	Upper/Lower Body - (3 sets) Power - 4 reps
Friday	Run 30 min @ 7.0 min/mile	None

WEEK 14	Endurance	Strength
Monday	Run 58 min @ 8.5 min/mile	None
Tuesday	Intervals: 6x 400 m sprints in 2 min (3:30 min rest)	Upper/Lower Body - (3 sets)
Wednesday	Run 40 min @ 8.0 min/mile	None
Thursday	None	Upper/Lower Body - (3 sets) Power - 4 reps
Friday	Run 30 min @ 7.0 min/mile	None

WEEK 15	Endurance	Strength
Monday	Run 30 min (easy pace)	Upper/Lower - (1 set)
Tuesday	None	None
Wednesday	Run 20 min (easy pace)	None
Thursday	None	None
Friday	Run 20 min (easy pace)	Upper/Lower - (1 set)

WEEK 16	Endurance	Strength
Monday	Run 30 min (easy pace)	Upper/Lower - (1 set)
Tuesday	None	None
Wednesday	Run 20 min (easy pace)	None
Thursday	None	None
Friday	Run 20 min (easy pace)	Upper/Lower - (1 set)

WEEK 17	Endurance	Strength
Monday	Run 30 min (easy pace)	None
Tuesday	None	Lower Body - (2 sets) Power - 2 sets
Wednesday	Run 20 min (easy pace)	None
Thursday	None	Lower Body - (2 sets) Power - 2 sets or Obstacle Course
Friday	Run 15 min (easy pace)	None

WEEK 18	Endurance	Strength
Monday	Team Sport	None
Tuesday	None	Lower Body - (1 set) Power - 1 set
Wednesday	Team Sport	None
Thursday	None	Lower Body - (1 set) Power - 1 set or Obstacle Course
Friday	Team Sport	None

WEEK 19	Endurance	Strength
Monday	Team Sport	None
Tuesday	None	None
Wednesday	Team Sport	None
Thursday	None	None
Friday	Team Sport	Obstacle Course

WEEK 20	Endurance	Strength
Monday	None	Upper/Lower - (2 sets)
Tuesday	Team Sports	None
Wednesday	None	Upper/Lower - (2 sets)
Thursday	Team Sports	Obstacle Course
Friday	None	Upper/Lower - (2 sets)

On week 21, begin the training cycle again, starting at week 1.

The 20-week PT cycle uses a modified form of periodization training. No athlete can maintain a continuous competitive edge - any attempt to do so will result in an overtrained state where mental and physical abilities will deteriorate.

Interruptions in the training schedule are expected. As a general guideline, for each week during which scheduled PT is not done, the unit should go back 2 weeks from the last week to safely restart the program.

c. <u>5-Day/6-Week Cardio/Muscular Strength Conditioning Program</u>

MONDAY	TUESDAY	WEDNESDAY	THURSDAY	FRIDAY
WEEK 1				
PFT	Resistance Training (RT)	Aerobic Training (AT) Ability Group Run (ABGR) 3-mi Upper Body Strength Training (UBS)	RT	Hike: 5-mi @ 2hrs/ w/30lbs
WEEK 2				
Circuit Training (CT)	RT	ABGR: 4-mi UBS	RT	Unit Run
WEEK 3				
ABGR: 4-mi	RT	ABGR: 4-mi	Anaerobic Training (ANT): 6x 100m sprints w/ 45-sec rest	Hike: 11km/6.8-mi <2.75hrs w/40lbs
WEEK 4				
RT	ABGR: 5-mi	CT	RT	Unit Run
WEEK 5				
ABGR: 5-mi UBS	RT	ABGR: 5-mi	ANT: 8x 100m sprints w/ 45-sec rest	Hike: 13km/8-mi <3.25hrs w/50lbs
WEEK 6				
PFT 4-mi	RT	ABGR: 6-mi UBS	RT	Unit Run

d. <u>Conditioning Program "A"</u>. Weekly aerobic program to address problems in the PFT and other distance running.

<u>Monday</u>: Distance/Time Run 30-minute run (slower pace than PFT run pace)

<u>Tuesday</u>: Cross training, Conduct cardiovascular training other than running; e.g., circuit training, swimming, or biking

<u>Wednesday</u>: Interval Training

Interval runs on track or known distance course - 4x 400 sprints;
- Increase # of sprints with conditioning
- Increase sprint distance with conditioning

<u>Thursday</u>: Cross Training

<u>Friday</u>: Distance Run - increase duration/speed from Monday's run

e. <u>Conditioning Program "B" (6 months)</u>. Program is designed to address PFT/distance running weaknesses, using a long-term (6 months) conditioning period. Run training should be conducted three times per week on non-consecutive days (e.g., Monday/Wednesday/Friday). One of the runs will be an interval run and the remaining two should be distance runs. The following schedule depicts the interval runs for every week for the 6-month period.

INTERVAL RUN TRAINING

WEEK	RUN DIST (YDS:MI)	MAX TIME (MIN:SEC)	WALK DIST: TIME (YDS:SEC)	REPS- RUN/WALK	TOTAL TIME (MIN:SEC)	TOTAL MILES
1,2	220: 1/8	1:00	110:60	11	22:00	2
3,4	330:	1:30	110:60	9	22:30	2.25
5,6	440: 1/4	2:20	110:75	8	26:40	2.5
7,8	660:	4:00	110:90	6	31:00	2.63
9,10	880: 1/2 440: 1/4	5:30 2:10	110:120 110:75	3 3	32:45	2.63
11,12	880: 1/2 440: 1/4	5:20 2:00	110:120 110:75	4 2	35:50	2.81
13,14	880: 1/2 660:	5:20 3:45	110:105 110:75	4 2	38:20	3.06
15,16	880: 1/2 660:	5:00 3:30	110:105 110:75	5 1	36:00	3.19
17,18	880: 1/2	4:45	110:90	6	37:30	3.38
19,20	1 M 4 laps 880: 1/2	9:00 4:30	110:120 110:90	1 4	35:00	3.31
21-24	1 M 4 laps 880: 1/2	8:30 4:15	110:105 110:90	1 4	33:15	3.31

The following chart shows the distance runs (per training session) for every week for the 6-month period. The distance runs should be conducted at 75% of maximum effort.

DISTANCE RUN TABLE

WEEKS	RUN DISTANCE (MILES)
1-4	2.5
5-8	2.75
9-12	3.0
13-16	3.25
17-20	3.5
21-24	4.0

Each PT session should include a substantial warmup, stretching, and cool-down from the Daily 16.

f. Conditioning Program "C" (Single Activity). The "single activity" conditioning schedule provides a 60-minute training session in which the single activity system is used with different events each day. In this program, 5 hours of training are scheduled for the week, and supplemented with integrated unit training each week. A bimonthly conditioning march of 10 miles in 3 hours is suggested to round out this program. Such a program could be used as a transition period (1-4 weeks) between other remedial/conditioning programs.

DAY	TIME	ACTIVITY
Monday	60-min	Running Activity
Tuesday	60-min	Combatives
Wednesday	60-min	Conditioning & Log Drills
Thursday	60-min	Conditioning & Relays
Friday	60-min	Running & Conditioning
Saturday (optional)		Running Activity

g. Conditioning Program "D" (Single Activity). This conditioning program schedule illustrates the use of another "single activity" system for 5.25 hours of training, to include sustaining-type activities for a unit that has passed through both the toughening and slow improvement stages of conditioning. The use of competitive activities is featured to hold interest and provide self-motivation. A bimonthly 10-mile/3-hour march under load will be necessary to round out this program.

DAY	TIME	ACTIVITY
Monday	45-min	Running - Terrain Running
Tuesday	75-min	Team Sports or Ability Group Run/Interval Sets
Wednesday	45-min	Running - Fartlek or Distance Run
Thursday	20-min 40-min	Log Drills Dual Combatives
Friday	90-min	Speed March
Saturday (optional)		Team Sports/Team Contests

h. Conditioning Program "E" (Multi-Activities). This program provides a more comprehensive training regimen of various training periods and activities, and the use of three systems during the training week. In this schedule, 3.5 hours are included for the week.

DAY	TIME	ACTIVITY
Monday	20-min 20-min 20-min	Rifle Drill Combatives Running
Tuesday	30-min	Daily 16 Grass Drills
Wednesday	30-min	Rifle Drill
Thursday	30-min	Circuit Interval Training
Friday	20-min 20-min 20-min	Strength Circuit Conditioning Drills Cross-Country Run
Saturday (optional)	60-min	Warmup Run/Grass Drills

On Monday, a 60-min period is available for PT with three, 20-min segments. Groups will progress through each scheduled activity in turn. On Tuesday, Wednesday, and Thursday, only 30-min periods are available, conducted by the entire unit. As with the Monday schedule, on Friday, the unit should conduct the 20-min segments as subgroups, again progressing through each scheduled activity in turn. Bimonthly conditioning marches will be necessary to augment the program.

APPENDIX D

BODY COMPOSITION PROGRAM (BCP) ASSIGNMENT PROCESS

LETTERHEAD

6100
Office Code
Date

From: Commanding Officer
To: Medical Officer

Subj: BODY COMPOSITION PROGRAM (BCP) EVALUATION ON RANK SNM
XXX XX XXX/XXXX

Ref: (a) MCO P6100.12

1. It has been determined that (RANK SNM)______________________________ is not within the Corps' body composition standards.

2. During the semi-annual body composition evaluation conducted on (YYMMDD) ____________________, SNM was found to weigh ________ pounds and has a ________ percent body fat. Per the reference, the Marine's maximum weight is ________ pounds based on a height of ________ inches.

3. SNM received a Physical Fitness Test (PFT) score on (YYMMDD) ________________, which is a ($1^{st}/2^{nd}/3^{rd}$) ________ Class score.

4. SNM was screened and does not meet the (PFT performance/percent body fat) __________________________ criteria and is therefore not eligible for the Physical Performance Evaluation.

5. SNM has been advised that the loss of ______ pounds or _____ percent body fat is required and this reduced weight/body fat percentage must be maintained within the 6-month BCP assignment period.

6. Request a medical evaluation, per the reference, be conducted to determine SNM's medical status for BCP and Remedial Physical Conditioning Program (RPCP) participation.

____
CO's Signature

Copy to:

APPENDIX D

BODY COMPOSITION PROGRAM (BCP) ASSIGNMENT PROCESS
LETTERHEAD

6100
Office Code
Date

FIRST ENDORSEMENT on (identify letter)

From: Medical Officer
To: Commanding Officer

Subj: BODY COMPOSITION PROGRAM (BCP) EVALUATION ON RANK SNM
XXX XX XXXX/XXXX

Ref: (a) MCO P6100.12

1. Returned.

2. Per the reference, it has been determined that (RANK SNM) ____________________ is not within body composition standards for the following (initial one of the following):

a. (_____) Due to an underlying cause or associated disease. Attached are the recommended follow-on medical tests/evaluations.

b. (_____) Due to an underlying cause or associated disease. Attached is the prescribed treatment for the condition. The Residential Obesity Program is/is not recommended for this individual.

c. (_____) Not to an underlying cause or associated disease. SNM is fit for participation in a BCP and Remedial Physical Conditioning Program (RPCP). Attached are the prescribed weight and body fat reduction goals and dietary guidelines.

d. (_____) Not to an underlying cause or associated disease. SNM is fit for participation in a BCP and alternative RPCP. Attached are the physical restrictions/limitations imposed, prognosis for return to a full duty status, weight and body fat reduction goals, and dietary guidelines.

____
MO's (ACHCP's) Signature

____
Grade

____
Branch of Service

____
Medical Title

____
Organization

APPENDIX D

BODY COMPOSITION PROGRAM (BCP) ASSIGNMENT PROCESS
LETTERHEAD

6100
Office Code
Date

SECOND ENDORSEMENT on (identify letter)

From: Commanding Officer
To: Individual Marine

Subj: BODY COMPOSITION PROGRAM (BCP) EVALUATION ON RANK SNM XXX XX XXXX/XXXX

Ref: (a) MCO P6100.12
(b) MCO P1080.40C
(c) MCO P1900.16F

1. Readdressed and forwarded.

2. It has been determined that you are not in adherence with Marine Corps body composition standards and are not eligible for the Physical Performance Evaluation. You are being assigned to a 6-month BCP as a ($1^{st}/2^{nd}$) ________ assignment. You are directed to meet the following weight reduction goal of ______ pounds or a body fat reduction goal of ______ percent per month. Your goal is to lose ______ pounds and ______ percent body fat.

3. Your mandatory participation in the BCP and following Program begins on (YYMMDD) _______________ and ends on (YYMMDD) _______________.

a. (_____) The Unit's Remedial Physical Conditioning Program (RPCP) for 6 months per reference (a).

b. (_____) The alternate RPCP as a result of physical restrictions/ limitations imposed by the Medical Officer per reference (a).

4. You are advised that assistance in meeting and maintaining your weight and body fat reduction goals is available through

____

____

____.

5. Official assignment to the 6-month BCP will be reported on the Unit Diary per reference (b).

6. You are advised that failure to comply with reference (a) may result in administrative separation per paragraph 6206 or 6215 of reference (c) respectively.

____
CO's Signature

____
Unit Diary Number

____
Date

APPENDIX D

BODY COMPOSITION PROGRAM (BCP) ASSIGNMENT PROCESS
LETTERHEAD

6100
Office Code
Date

THIRD ENDORSEMENT on (identify letter)

From: Individual Marine
To: Commanding Officer

Subj: BODY COMPOSITION PROGRAM (BCP) EVALUATION ON RANK SNM XXX XX XXXX/XXXX

1. Returned with acknowledgments. (SNM must read and initial all paragraphs below.)

2. (_____) I understand I am not in adherence with Marine Corps body composition standards, and it is not due to an underlying cause or associated disease, based on the medical evaluation completed by a Medical Officer (MO).

3. (_____) I understand I am not eligible for the Physical Performance Evaluation and I am being assigned to the BCP, as a ($1^{st}/2^{nd}$) ________ assignment.

4. (_____) I understand I am being directed to meet the following weight reduction goal of ______ pounds or a body fat reduction goal of ______ percent per month, and my weight goal is ______ pounds and body fat goal is ______ percent.

5. I understand my responsibility to (initial all):

a. (_____) Participate in the Unit's BCP for a total of 6 months.

b. (_____) Participate in the Unit Remedial Physical Conditioning Program (RPCP) or alternate RPCP, as prescribed by the evaluating MO for the duration of my 6-month BCP assignment.

c. (_____) Have my weight and body fat measurements taken and recorded once every 2 weeks by the Command Physical Training Representative.

d. (_____) Follow the dietary regimen prescribed by the evaluating MO.

e. (_____) Complete the MCI Course 3316, "Basic Nutrition." (First assignment to BCP only).

____
Individual Marine's Signature

APPENDIX D

BODY COMPOSITION PROGRAM (BCP) ASSIGNMENT PROCESS
LETTERHEAD

6100
Office Code
Date

FOURTH ENDORSEMENT on (identify letter)

From: Commanding Officer
To: Medical Officer

Subj: BODY COMPOSITION PROGRAM (BCP) EVALUATION ON RANK SNM
XXX XX XXXX/XXXX

1. Readdressed and forwarded for action.

2. SNM has been participating in the BCP and Remedial Physical Conditioning Program for 4 months. A medical reevaluation is requested on SNM, based on progress made towards weight and body fat reduction goals. If weight and body fat reduction goals are not being met or maintained, verify the diagnosis of no underlying cause or associated disease.

3. SNM has lost ______ pound(s) and ______ percent body fat, since being placed to the BCP on (YYMMDD) ______________.

____
CO's Signature

APPENDIX D

BODY COMPOSITION PROGRAM (BCP) ASSIGNMENT PROCESS
LETTERHEAD

6100
Office Code
Date

FIFTH ENDORSEMENT on (identify letter)

From: Medical Officer
To: Commanding Officer

Subj: BODY COMPOSITION PROGRAM (BCP) EVALUATION ON RANK SNM
XXX XX XXXX/XXXX

Ref: (a) MCO P6100.12

1. Returned.

2. SNM has been reevaluated, resulting in the following diagnosis/ recommendation: (Initial appropriate paragraph)

a. (_____) SNM's current body composition status is due to an underlying cause or associated disease. Attached are the diagnosis and recommended follow-on actions to be taken per paragraph 3200 of the reference.

b. (_____) SNM's present body composition status is not due to an underlying cause or associated disease. Participation in the BCP and Remedial Physical Conditioning Program should continue until Program assignment expires.

3. Based on the above diagnoses, medical comments and recommendations are attached.

____
MO's (ACHCP's) Signature

____
Grade

____
Branch of Service

____
Medical Title

____
Organization

APPENDIX D

BODY COMPOSITION PROGRAM (BCP) ASSIGNMENT PROCESS
LETTERHEAD

6100
Office Code
Date

SIXTH ENDORSEMENT on (identify letter)

From: Commanding Officer
To: Individual Marine

Subj: BODY COMPOSITION PROGRAM (BCP) EVALUATION ON RANK SNM
XXX XX XXXX/XXXX

Ref: (a) MCO P1080.40C
(b) MCO P1900.16F

1. Readdressed and forwarded.

2. Your ($1^{st}/2^{nd}$) ________ 6-month assignment to the BCP expired on (YYMMDD) ________________.

3. Based on your progress while assigned to the BCP, as documented in your BCP file, it has been determined that: (Initial one)

a. (_____) You have attained and maintained the Marine Corps' body composition standards, and will be officially removed from the BCP (and Remedial Physical Conditioning Program (RPCP)) on the Unit Diary per reference (a).

(1) (_____) FIRST ASSIGNMENT PARTICIPANTS. If at any time during your Marine Corps career you fail to meet the body composition standards a second time, you may be granted a second BCP (and RPCP) assignment of 6 months with no extension.

(2) (_____) SECOND ASSIGNMENT PARTICIPANTS. If at any time during your Marine Corps career you fail to meet the body composition standards a third time, you will automatically be processed for administrative separation per reference (b).

b. (_____) FIRST ASSIGNMENT PARTICIPANTS ONLY. You have not met the Marine Corps' body composition standards, but have made satisfactory progress during your first 6-month BCP assignment. You have been granted an extension for ______ additional months, and will be required to return to body composition standards upon Program extension expiration on (YYMMDD) _______________. This extension has been entered on the Unit Diary per reference (a).

(1) You will continue your participation in the BCP and RPCP throughout the extension period.

(2) Thirty-days prior to Program expiration, you will be medically reevaluated.

(3) Upon Program expiration, if body composition standards are met, you will be officially removed from the BCP (and RPCP). If body composition standards are not met, you will be processed for administrative separation per paragraph 6215 in reference (b).

BODY COMPOSITION PROGRAM (BCP) ASSIGNMENT PROCESS

SIXTH ENDORSEMENT on (identify letter)

Subj: BODY COMPOSITION PROGRAM (BCP) EVALUATION ON RANK SNM
XXX XX XXXX/XXXX

c. (_____) You have failed to meet the Marine Corps' body composition standards and will receive a 6105 counseling entry on your SRB/OQR's Page 11, and be processed for administrative separation per paragraph 6206 or 6215 in reference (b) respectively.

____
CO's Signature

____
Individual Marine's Signature

____
Unit Diary Number

____
Date

2

LETTERHEAD

6100
Office Code
Date

SEVENTH ENDORSEMENT on (identify letter)

From: Commanding Officer
To: Individual Marine

Subj: BODY COMPOSITION PROGRAM (BCP) EVALUATION ON RANK SNM
XXX XX XXXX/XXXX

Ref: (a) MCO P1080.40C
(b) MCO P1900.16F

1. Readdressed and forwarded.

2. Your extension to your first 6-month BCP assignment expired on (YYMMDD) ______________.

3. Based on your progress while assigned to the BCP, as documented in your BCP file, it has been determined that: (Initial one)

a. (_____) You have attained and maintained the Marine Corps' body composition standards, and will be officially removed from the BCP (and Remedial Physical Conditioning Program (RPCP)) on the Unit Diary per reference (a). If at any time during your Marine Corps career you fail to meet the body composition standards a second time, you may be granted a second BCP (and RPCP) assignment of 6-months with no extension.

b. (_____) You have failed to meet the Marine Corps' body composition standards and will receive a 6105 counseling entry on your SRB/OQR's Page 11, and be processed for administrative separation per paragraph 6206 or 6215 in reference (b) respectively.

____
CO's Signature

____
Individual Marine's Signature

____
Unit Diary Number

____
Date

APPENDIX D

PHYSICAL PERFORMANCE EVALUATION FLOWCHART

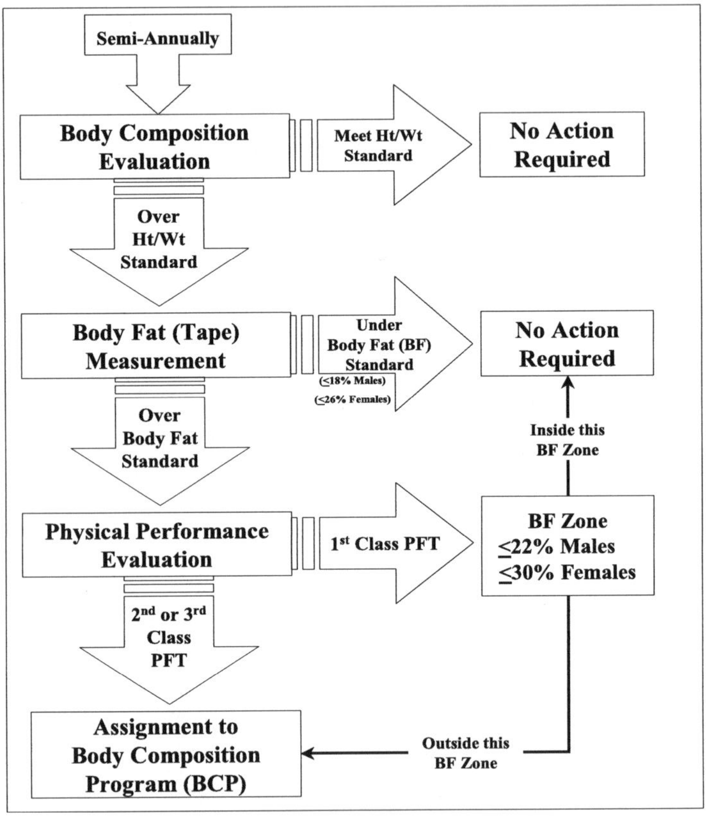

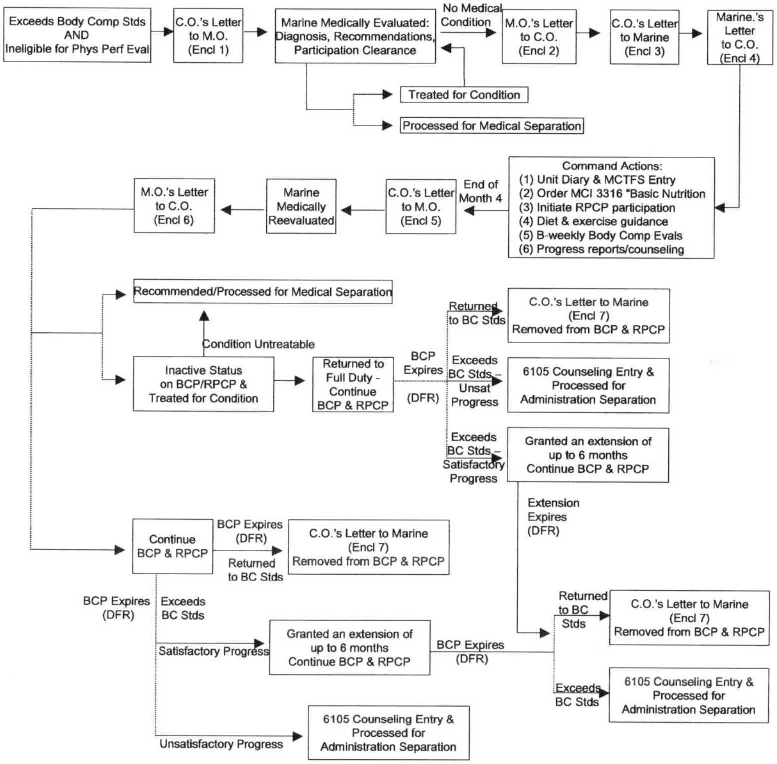

Appendix D: Body Composition Program Assignment Process

NOTES:

(1) Marines assigned to a Body Composition Program (BCP) will also be assigned to the Remedial Physical Conditioning Program (RPCP) for the duration of the 6-month BCP Assignment/Extension.
(2) A Marine on light/limited duty will participate in an alternative RPCP.
(3) 2nd-Assignment Marines, who successfully completed their 1st Assignment and fail to meet BC standards later in their career, will be given 6 months (with no estension) for a 2nd Assignment.
(4) No Third Chances: Marines failing to meet BC standards a third time in their career will be immediately processed for administrative separation, and will informally participate in the BCP & RPCP until discharged. No 3rd Assignment will be given under any circumstances.
(5) Marines will receive a 6105 counseling entry on their SRB/OQR's Page 11 per paragraph 6206 or 6215 of MCO P1900.16F respectively.

APPENDIX E

AGE 46 AND OVER RISK FACTOR SCREENING CHECKLIST

AGE 46 AND OVER RISK FACTOR SCREENING

NAME (LAST, FIRST, M.I.):		SSN:
COMMAND:	SECTION:	PHONE:

SECTION A: (COMPLETED BY THE MARINE)	Circle
(1) ARE YOU NOT IN ACCORDANCE WITH MEDICAL PHYSICAL EXAMINATION REQUIREMENTS FOR YOUR AGE?	YES NO
(2) DATE OF LAST PERIODIC PHYSICAL EXAMINATION:	YES NO
SECTION B: (COMPLETED BY THE MARINE)	
(1) DO YOU HAVE A HISTORY OF HEART DISEASE OR HIGH BLOOD PRESSURE WHICH REQUIRES YOU TO RESTRICT PHYSICAL ACTIVITY OR SEEK MEDICAL TREATMENT?	YES NO
(2) SINCE YOUR LAST PERIODIC PHYSICAL EXAMINATION, SEMI-ANNUAL RISK FACTOR SCREENING, OR PFT, HAS THERE BEEN:	
A. A CHANGE IN YOUR HISTORY OF HEART DISEASE OR HIGH BLOOD PRESSURE WHICH REQUIRED YOU TO RESTRICT PHYSICAL ACTIVITY OR SEEK MEDICAL TREATMENT?	YES NO
B. A CHANGE IN YOUR FAMILY HISTORY SUCH AS IMMEDIATE FAMILY MEMBERS EXPERIENCING A HEART ATTACK OR DIED OF HEART DISEASE BEFORE THEY WERE 45 YEARS OLD?	YES NO
C. A CHANGE IN YOUR INCIDENCE OF DISCOMFORT TO YOUR CHEST, ARMS, OR NECK WHILE EXERCISING?	YES NO
D. AN INCIDENT WHERE YOU WERE PRONE TO HEAT EXHAUSTION/HEAT STROKE, FEELING FAINT, OR FEELING THAT YOU WERE ABOUT TO LOSE CONSCIOUSNESS?	YES NO
E. A CHANGE IN ANY MEDICAL CONDITION (E.G., DIABETES, ASTHMA, BONE OR JOINT DISEASE) WHICH YOU THINK MIGHT LIMIT YOUR PARTICIPATION IN THE PFT?	YES NO
F. A SIGNIFICANT WEIGHT CHANGE?	YES NO
G. A CHANGE IN YOUR SMOKING HABITS SUCH THAT YOU NOW SMOKE ONE OR MORE PACKS OF CIGARETTES DAILY?	YES NO
MARINE'S SIGNATURE : DATE:	
SECTION C: (COMPLETED BY THE MEDICAL OFFICER (M.O.) IF REQUIRED)	
DATE OF MEDICAL EVALUATION: ____________________	
(1) THE MARINE'S PERIODIC PHYSICAL EXAMINATION IS CURRENT (Y/N)? ________ REMARKS:____________________	
(2) THE MARINE ANSWERED 'YES' TO QUESTIONS IN SECTION (B)(Y/N)? ______ REMARKS:____________________	
SECTION D: (COMPLETED BY THE MEDICAL OFFICER (M.O.) IF REQUIRED)	
(1) MEDICAL RECOMMENDATION: REFERRED FOR FURTHER MEDICAL CONSULTATION (Y/N)? _____ (E.G., PSYCHOLOGY, NUTRITION, INTERNAL MEDICINE, PHYSICAL THERAPY) REMARKS:____________________	
(2) CLEARED FOR PARTICIPATION IN PFT (Y/N)? _____ REMARKS:____________________	
SIGNATURE OF M.O. DATE:	
PRINTED NAME/RANK/COMMAND/UNIT/PHONE:	

APPENDIX F

PHYSICAL FITNESS TEST AND BODY COMPOSITION WORKSHEET

UNIT:

DATE:

MONITOR:

Remarks

* %BF required ONLY if Marine exceeds Ht/Wt standards.

** Semiannual Risk Factor Screening required ONLY if Marine is age 46 and over.

NAME/SSN	RANK	AGE	GENDER M/F	HT (In.)	WT (lbs.)	% BF*	Risk Factor Screening**	Pullups Flx Arm Hang	SCORE	Crunches	SCORE	3-Mile Run	SCORE	PFT TOTAL	Partial PFT TOTAL	CLASS	Phys Perf Eval (Y/N)

APPENDIX G

PHYSICAL FITNESS TEST SCORING TABLE

Sec I: FEMALES

Points	Flexed-Arm Hang	Crunches	3-Mile Run	Points	Flexed-Arm Hang	Crunches	3-Mile Run
100	70 sec	100	21:00	**50**	45 sec	50	29:20
99		99	21:10	**49**		49	29:30
98	69 sec	98	21:20	**48**	44 sec	48	29:40
97		97	21:30	**47**		47	29:50
96	68 sec	96	21:40	**46**	43 sec	46	30:00
95		95	21:50	**45**		45	30:10
94	67 sec	94	22:00	**44**	42 sec	44	30:20
93		93	22:10	**43**		43	30:30
92	66 sec	92	22:20	**42**	41 sec	42	30:40
91		91	22:30	**41**		41	30:50
90	65 sec	90	22:40	**40**	40 sec	40	31:00
89		89	22:50	**39**	39 sec	x	31:10
88	64 sec	88	23:00	**38**	38 sec	x	31:20
87		87	23:10	**37**	37 sec	x	31:30
86	63 sec	86	23:20	**36**	36 sec	x	31:40
85		85	23:30	**35**	35 sec	x	31:50
84	62 sec	84	23:40	**34**	34 sec	x	32:00
83		83	23:50	**33**	33 sec	x	32:10
82	61 sec	82	24:00	**32**	32 sec	x	32:20
81		81	24:10	**31**	31 sec	x	32:30
80	60 sec	80	24:20	**30**	30 sec	x	32:40
79		79	24:30	**29**	29 sec	x	32:50
78	59 sec	78	24:40	**28**	28 sec	x	33:00
77		77	24:50	**27**	27 sec	x	33:10
76	58 sec	76	25:00	**26**	26 sec	x	33:20
75		75	25:10	**25**	25 sec	x	33:30
74	57 sec	74	25:20	**24**	24 sec	x	33:40
73		73	25:30	**23**	23 sec	x	33:50
72	56 sec	72	25:40	**22**	22 sec	x	34:00
71		71	25:50	**21**	21 sec	x	34:10
70	55 sec	70	26:00	**20**	20 sec	x	34:20
69		69	26:10	**19**	19 sec	x	34:30
68	54 sec	68	26:20	**18**	18 sec	x	34:40
67		67	26:30	**17**	17 sec	x	34:50
66	53 sec	66	26:40	**16**	16 sec	x	35:00
65		65	26:50	**15**	15 sec	x	35:10
64	52 sec	64	27:00	**14**	x	x	35:20
63		63	27:10	**13**	x	x	35:30
62	51 sec	62	27:20	**12**	x	x	35:40
61		61	27:30	**11**	x	x	35:50
60	50 sec	60	27:40	**10**	x	x	36:00
59		59	27:50	**9**	x	x	x
58	49 sec	58	28:00	**8**	x	x	x
57		57	28:10	**7**	x	x	x
56	48 sec	56	28:20	**6**	x	x	x
55		55	28:30	**5**	x	x	x
54	47 sec	54	28:40	**4**	x	x	x
53		53	28:50	**3**	x	x	x
52	46 sec	52	29:00	**2**	x	x	x
51		51	29:10	**1**	x	x	x

***Round up all values (e.g., 21:01 to 21:09 equals 99 points)**

APPENDIX G

PHYSICAL FITNESS TEST SCORING TABLE

Sec II:	**MALES**						
Points	Pull-ups	Crunches	3-Mile Run	**Points**	Pull-ups	Crunches	3-Mile Run
100	20	100	18:00	**50**	10	50	26:20
99		99	18:10	**49**		49	26:30
98		98	18:20	**48**		48	26:40
97		97	18:30	**47**		47	26:50
96		96	18:40	**46**		46	27:00
95	19	95	18:50	**45**	9	45	27:10
94		94	19:00	**44**		44	27:20
93		93	19:10	**43**		43	27:30
92		92	19:20	**42**		42	27:40
91		91	19:30	**41**		41	27:50
90	18	90	19:40	**40**	8	40	28:00
89		89	19:50	**39**		x	28:10
88		88	20:00	**38**		x	28:20
87		87	20:10	**37**		x	28:30
86		86	20:20	**36**		x	28:40
85	17	85	20:30	**35**	7	x	28:50
84		84	20:40	**34**		x	29:00
83		83	20:50	**33**		x	29:10
82		82	21:00	**32**		x	29:20
81		81	21:10	**31**		x	29:30
80	16	80	21:20	**30**	6	x	29:40
79		79	21:30	**29**		x	29:50
78		78	21:40	**28**		x	30:00
77		77	21:50	**27**		x	30:10
76		76	22:00	**26**		x	30:20
75	15	75	22:10	**25**	5	x	30:30
74		74	22:20	**24**		x	30:40
73		73	22:30	**23**		x	30:50
72		72	22:40	**22**		x	31:00
71		71	22:50	**21**		x	31:10
70	14	70	23:00	**20**	4	x	31:20
69		69	23:10	**19**		x	31:30
68		68	23:20	**18**		x	31:40
67		67	23:30	**17**		x	31:50
66		66	23:40	**16**		x	32:00
65	13	65	23:50	**15**	3	x	32:10
64		64	24:00	**14**	x	x	32:20
63		63	24:10	**13**	x	x	32:30
62		62	24:20	**12**	x	x	32:40
61		61	24:30	**11**	x	x	32:50
60	12	60	24:40	**10**	x	x	33:00
59		59	24:50	**9**	x	x	x
58		58	25:00	**8**	x	x	x
57		57	25:10	**7**	x	x	x
56		56	25:20	**6**	x	x	x
55	11	55	25:30	**5**	x	x	x
54		54	25:40	**4**	x	x	x
53		53	25:50	**3**	x	x	x
52		52	26:00	**2**	x	X	x
51		51	26:10	**1**	x	x	x

*** Round up all values (e.g., 18:01 to 18:09 equals 99 points)**

APPENDIX H

ALTITUDE COMPENSATION FOR 3.0 MILE PHYSICAL FITNESS RUN CHART

Points	3.0-Mile (Male)	3.0-Mile (Female)	Points	3.0-Mile (Male)	3.0-Mile (Female)	Points	3.0-Mile (Male)	3.0-Mile (Female)
100	19:30	22:30	**65**	25:20	:20	**30**	31:10	34:10
99	199:40	22:40	**64**	25:30	28:30	**29**	31:20	34:20
98	19:50	22:50	**63**	25:40	28:40	**28**	31:30	34:30
97	20:00	23:00	**62**	25:50	28:50	**27**	31:40	34:40
96	20:10	23:10	**61**	26:00	29:00	**26**	31:50	34:50
95	20:20	23:20	**60**	26:10	29:10	**25**	32:00	35:00
94	20:30	23:30	**59**	26:20	29:20	**24**	32:10	35:10
93	20:40	23:40	**58**	26:30	29:30	**23**	32:20	35:20
92	20:50	23:50	**57**	26:40	29:40	**22**	32:30	35:30
91	21:00	24:00	**56**	26:50	29:50	**21**	32:40	35:40
90	21:10	24:10	**55**	27:00	30:00	**20**	32:50	35:50
89	21:20	24:20	**54**	27:10	30:10	**19**	33:00	36:00
88	21:30	24:30	**53**	27:20	30:20	**18**	33:10	36:10
87	21:40	24:40	**52**	27:30	30:30	**17**	33:20	36:20
86	21:50	24:50	**51**	27:40	30:40	**16**	33:30	36:30
85	22:00	25:00	**50**	27:50	30:50	**15**	33:40	36:40
84	22:10	25:10	**49**	28:00	31:00	**14**	33:50	36:50
83	22:20	25:20	**48**	28:10	31:10	**13**	34:00	37:00
82	22:30	25:30	**47**	28:20	31:20	**12**	34:10	37:10
81	22:40	25:40	**46**	28:30	31:30	**11**	34:20	37:20
80	22:50	25:50	**45**	28:40	31:40	**10**	34:30	37:30
79	23:00	26:00	**44**	28:50	31:50	**9**	34:40	37:40
78	23:10	26:10	**43**	29:00	32:00	**8**	34:50	37:50
77	23:20	26:20	**42**	29:10	32:10	**7**	35:00	38:00
76	23:30	26:30	**41**	29:20	32:20	**6**	35:10	38:10
75	23:40	26:40	**40**	29:30	32:30	**5**	35:20	38:20
74	23:50	26:50	**39**	29:40	32:40	**4**	35:30	38:30
73	24:00	27:00	**38**	29:50	32:50	**3**	35:40	38:40
72	24:10	27:10	**37**	30:00	33:00	**2**	35:50	38:50
71	24:20	27:20	**36**	30:10	33:10	**1**	36:00	39:00
70	24:30	27:30	**35**	30:20	33:20			
69	24:40	27:40	**34**	30:30	33:30			
68	24:50	27:50	**33**	30:40	33:40			
67	25:00	28:00	**32**	30:50	33:50			
66	25:10	28:10	**31**	31:00	34:00			

APPENDIX I

SECTION 1: HEIGHT/WEIGHT STANDARDS

MALES

HEIGHT (Inches)	Maximum Standard (Pounds)	Minimum Standard (Pounds)
58"	132	91
59"	136	94
60"	141	97
61"	146	100
62"	150	104
63"	155	107
64"	160	110
65"	165	114
66"	170	117
67"	176	121
68"	181	125
69"	186	128
70"	192	132
71"	197	136
72"	203	140
73"	208	144
74"	214	148
75"	220	152
76"	226	156
77"	232	160
78"	238	164
79"	244	168
80"	250	173

Max BF%:	18%

FEMALES

HEIGHT (Inches)	Maximum Standard (Pounds)	Minimum Standard (Pounds)
58"	120	91
59"	124	94
60"	128	97
61"	132	100
62"	137	104
63"	141	107
64"	146	110
65"	150	114
66"	155	117
67"	160	121
68"	164	125
69"	169	128
70"	174	132
71"	179	136
72"	184	140
73"	189	144
74"	195	148
75"	200	152
76"	205	156
77"	211	160
78"	216	164
79"	222	168
80"	228	173

Max BF%:	26%

APPENDIX I

SECTION 2: PERCENT BODY FAT ESTIMATION

PERCENT BODY FAT ESTIMATION FOR MEN

Circumference Value*	Height (in)									
	60.0	60.5	61.0	61.5	62.0	62.5	63.0	63.5	64.0	64.5
13.5	9	9								
14.0	11	11	10	10	10	10	9	9		
14.5	12	12	12	11	11	11	11	10	10	10
15.0	13	13	13	13	12	12	12	12	11	11
15.5	15	14	14	14	14	13	13	13	13	12
16.0	16	16	15	15	15	15	14	14	14	14
16.5	17	17	16	16	16	16	15	15	15	15
17.0	18	18	18	17	17	17	17	16	16	16
17.5	19	19	19	18	18	18	18	17	17	17
18.0	20	20	20	19	19	19	19	18	18	18
18.5	21	21	21	20	20	20	20	19	19	19
19.0	22	22	22	21	21	21	21	20	20	20
19.5	23	23	23	22	22	22	22	21	21	21
20.0	24	24	24	23	23	23	23	22	22	22
20.5	25	25	25	24	24	24	24	23	23	23
21.0	26	26	25	25	25	25	24	24	24	24
21.5	27	27	26	26	26	26	25	25	25	25
22.0	28	27	27	27	27	26	26	26	26	25
22.5	29	28	28	28	28	27	27	27	27	26
23.0	29	29	29	29	28	28	28	28	27	27
23.5	30	30	30	29	29	29	29	28	28	28
24.0	31	31	30	30	30	30	29	29	29	29
24.5	32	31	31	31	31	30	30	30	30	29
25.0	32	32	32	32	31	31	31	31	30	30
25.5	33	33	33	32	32	32	32	31	31	31
26.0	34	34	33	33	33	33	32	32	32	32
26.5	35	34	34	34	34	33	33	33	33	32
27.0	35	35	35	35	34	34	34	34	33	33
27.5	36	36	36	35	35	35	35	34	34	34
28.0	37	36	36	36	36	35	35	35	35	34
28.5			37	37	36	36	36	36	35	35
29.0					37	37	37	36	36	36
29.5								37	37	36
30.0										
30.5										
31.0										
31.5										
32.0										
32.5										
33.0										
33.5										
34.0										
34.5										
35.0										

* Circumference Value = abdomen circumference - neck circumference (in inches)

PERCENT BODY FAT ESTIMATION FOR MEN

Circumference Value*	Height (in)									
	65.0	65.5	66.0	66.5	67.0	67.5	68.0	68.5	69.0	69.5
13.5										
14.0										
14.5	10	9	9							
15.0	11	11	10	10	10	10	10	9	9	
15.5	12	12	12	11	11	11	11	11	10	10
16.0	13	13	13	13	12	12	12	12	12	11
16.5	14	14	14	14	14	13	13	13	13	12
17.0	16	15	15	15	15	14	14	14	14	14
17.5	17	16	16	16	16	16	15	15	15	15
18.0	18	18	17	17	17	17	16	16	16	16
18.5	19	19	18	18	18	18	17	17	17	17
19.0	20	20	19	19	19	19	18	18	18	18
19.5	21	21	20	20	20	20	19	19	19	19
20.0	22	21	21	21	21	21	20	20	20	20
20.5	23	22	22	22	22	21	21	21	21	21
21.0	24	23	23	23	23	22	22	22	22	21
21.5	24	24	24	24	23	23	23	23	23	22
22.0	25	25	25	25	24	24	24	24	23	23
22.5	26	26	26	25	25	25	25	24	24	24
23.0	27	27	26	26	26	26	26	25	25	25
23.5	28	27	27	27	27	27	26	26	26	26
24.0	28	28	28	28	28	27	27	27	27	26
24.5	29	29	29	29	28	28	28	28	27	27
25.0	30	30	30	29	29	29	29	28	28	28
25.5	31	31	30	30	30	30	29	29	29	29
26.0	31	31	31	31	31	30	30	30	30	29
26.5	32	32	32	32	31	31	31	31	30	30
27.0	33	33	32	32	32	32	32	31	31	31
27.5	34	33	33	33	33	32	32	32	32	32
28.0	34	34	34	34	33	33	33	33	32	32
28.5	35	35	34	34	34	34	34	33	33	33
29.0	36	35	35	35	35	34	34	34	34	34
29.5	36	36	36	36	35	35	35	35	34	34
30.0	37	37	36	36	36	36	35	35	35	35
30.5			37	37	37	36	36	36	36	35
31.0						37	37	36	36	36
31.5									37	37
32.0										
32.5										
33.0										
33.5										
34.0										
34.5										
35.0										

* Circumference Value = abdomen circumference - neck circumference (in inches)

PERCENT BODY FAT ESTIMATION FOR MEN

Circumference Value*	Height (in)									
	70.0	70.5	71.0	71.5	72.0	72.5	73.0	73.5	74.0	74.5
13.5										
14.0										
14.5										
15.0										
15.5	10	10	9	9	9					
16.0	11	11	11	10	10	10	10	10	9	9
16.5	12	12	12	12	11	11	11	11	11	10
17.0	13	13	13	13	13	12	12	12	12	11
17.5	14	14	14	14	14	13	13	13	13	13
18.0	15	15	15	15	15	14	14	14	14	14
18.5	17	16	16	16	16	15	15	15	15	15
19.0	18	17	17	17	17	16	16	16	16	16
19.5	18	18	18	18	18	17	17	17	17	17
20.0	19	19	19	19	19	18	18	18	18	18
20.5	20	20	20	20	19	19	19	19	19	18
21.0	21	21	21	21	20	20	20	20	20	19
21.5	22	22	22	21	21	21	21	21	20	20
22.0	23	23	23	22	22	22	22	22	21	21
22.5	24	24	23	23	23	23	23	22	22	22
23.0	25	24	24	24	24	24	23	23	23	23
23.5	25	25	25	25	25	24	24	24	24	24
24.0	26	26	26	26	25	25	25	25	25	24
24.5	27	27	27	26	26	26	26	26	25	25
25.0	28	28	27	27	27	27	26	26	26	26
25.5	29	28	28	28	28	27	27	27	27	27
26.0	29	29	29	29	28	28	28	28	28	27
26.5	30	30	30	29	29	29	29	28	28	28
27.0	31	30	30	30	30	30	29	29	29	29
27.5	31	31	31	31	30	30	30	30	30	29
28.0	32	32	32	31	31	31	31	31	30	30
28.5	33	32	32	32	32	32	31	31	31	31
29.0	33	33	33	33	32	32	32	32	32	31
29.5	34	34	34	33	33	33	33	32	32	32
30.0	35	34	34	34	34	34	33	33	33	33
30.5	35	35	35	35	34	34	34	34	34	33
31.0	36	36	35	35	35	35	35	34	34	34
31.5	36	36	36	36	36	35	35	35	35	35
32.0	37	37	37	36	36	36	36	36	35	35
32.5				37	37	36	36	36	36	36
33.0							37	37	36	36
33.5										37
34.0										
34.5										
35.0										

* Circumference Value = abdomen circumference - neck circumference (in inches)

PERCENT BODY FAT ESTIMATION FOR MEN

Circumference Value*	Height (in)									
	75.0	75.5	76.0	76.5	77.0	77.5	78.0	78.5	79.0	79.5
13.5										
14.0										
14.5										
15.0										
15.5										
16.0										
16.5	10	10	10	10	9	9				
17.0	11	11	11	11	10	10	10	10	10	9
17.5	12	12	12	12	12	11	11	11	11	11
18.0	13	13	13	13	13	12	12	12	12	12
18.5	14	14	14	14	14	13	13	13	13	13
19.0	15	15	15	15	15	14	14	14	14	14
19.5	16	16	16	16	16	15	15	15	15	15
20.0	17	17	17	17	17	16	16	16	16	16
20.5	18	18	18	18	17	17	17	17	17	16
21.0	19	19	19	19	18	18	18	18	18	17
21.5	20	20	20	19	19	19	19	19	18	18
22.0	21	21	20	20	20	20	20	20	19	19
22.5	22	22	21	21	21	21	21	20	20	20
23.0	23	22	22	22	22	22	21	21	21	21
23.5	23	23	23	23	23	22	22	22	22	22
24.0	24	24	24	24	23	23	23	23	23	22
24.5	25	25	25	24	24	24	24	24	23	23
25.0	26	25	25	25	25	25	24	24	24	24
25.5	26	26	26	26	26	25	25	25	25	25
26.0	27	27	27	27	26	26	26	26	26	25
26.5	28	28	27	27	27	27	27	26	26	26
27.0	29	28	28	28	28	28	27	27	27	27
27.5	29	29	29	29	28	28	28	28	28	27
28.0	30	30	29	29	29	29	29	29	28	28
28.5	31	30	30	30	30	30	29	29	29	29
29.0	31	31	31	31	30	30	30	30	30	29
29.5	32	32	31	31	31	31	31	30	30	30
30.0	32	32	32	32	32	31	31	31	31	31
30.5	33	33	33	32	32	32	32	32	32	31
31.0	34	33	33	33	33	33	33	32	32	32
31.5	34	34	34	34	33	33	33	33	33	33
32.0	35	35	34	34	34	34	34	33	33	33
32.5	35	35	35	35	35	34	34	34	34	34
33.0	36	36	36	35	35	35	35	35	34	34
33.5	37	36	36	36	36	36	35	35	35	35
34.0		37	37	37	36	36	36	36	36	35
34.5					37	37	37	36	36	36
35.0								37	37	36

* Circumference Value = abdomen circumference - neck circumference (in inches)

PERCENT BODY FAT ESTIMATION FOR WOMEN

Circumference Value*	Height (in)									
	58.0	58.5	59.0	59.5	60.0	60.5	61.0	61.5	62.0	62.5
45.0	19									
45.5	20	20	19							
46.0	21	20	20	20	19					
46.5	21	21	21	20	20	20	19	19		
47.0	22	22	22	21	21	20	20	20	19	19
47.5	23	23	22	22	22	21	21	21	20	20
48.0	24	23	23	23	22	22	22	21	21	21
48.5	24	24	24	23	23	23	22	22	22	21
49.0	25	25	24	24	24	23	23	23	22	22
49.5	26	26	25	25	24	24	24	23	23	23
50.0	27	26	26	26	25	25	24	24	24	23
50.5	27	27	27	26	26	26	25	25	25	24
51.0	28	28	27	27	27	26	26	26	25	25
51.5	29	28	28	28	27	27	27	26	26	26
52.0	29	29	29	28	28	28	27	27	27	26
52.5	30	30	29	29	29	28	28	28	27	27
53.0	31	30	30	30	29	29	29	28	28	28
53.5	31	31	31	30	30	30	29	29	29	28
54.0	32	32	31	31	31	30	30	30	29	29
54.5	33	32	32	32	31	31	31	30	30	30
55.0	33	33	33	32	32	32	31	31	31	30
55.5	34	34	33	33	33	32	32	32	31	31
56.0	35	34	34	34	33	33	33	32	32	31
56.5	35	35	35	34	34	34	33	33	32	32
57.0	36	36	35	35	34	34	34	33	33	33
57.5	37	36	36	35	35	35	34	34	34	33
58.0	37	37	36	36	36	35	35	35	34	34
58.5	38	37	37	37	36	36	36	35	35	35
59.0	38	38	38	37	37	37	36	36	36	35
59.5	39	39	38	38	38	37	37	36	36	36
60.0	40	39	39	38	38	38	37	37	37	36
60.5	40	40	39	39	39	38	38	38	37	37
61.0	41	40	40	40	39	39	39	38	38	38
61.5	41	41	41	40	40	40	39	39	38	38
62.0	42	42	41	41	40	40	40	39	39	39

* Circumference Value = waist circumference + hip circumference - neck circumference (in inches)

PERCENT BODY FAT ESTIMATION FOR WOMEN

Circumference Value*	Height (in)									
	58.0	58.5	59.0	59.5	60.0	60.5	61.0	61.5	62.0	62.5
62.5	42	42	42	41	41	41	40	40	40	39
63.0	43	43	42	42	42	41	41	41	40	40
63.5	44	43	43	42	42	42	41	41	41	40
64.0	44	44	43	43	43	42	42	42	41	41
64.5	45	44	44	44	43	43	43	42	42	42
65.0	45	45	45	44	44	43	43	43	42	42
65.5	46	45	45	45	44	44	44	43	43	43
66.0	46	46	46	45	45	45	44	44	43	43
66.5	47	46	46	46	45	45	45	44	44	44
67.0			47	46	46	46	45	45	45	44
67.5				47	46	46	46	45	45	45
68.0					47	47	46	46	46	45
68.5							47	46	46	46
69.0								47	47	46
69.5										47
70.0										
70.5										
71.0										
71.5										
72.0										
72.5										
73.0										
73.5										
74.0										
74.5										
75.0										
75.5										
76.0										
76.5										
77.0										
77.5										
78.0										
78.5										
79.0										
79.5										

* Circumference Value = waist circumference + hip circumference - neck circumference (in inches)

SECTION 2: PERCENT BODY FAT ESTIMATION

PERCENT BODY FAT ESTIMATION FOR WOMEN

Circumference Value*	Height (in)									
	63.0	63.5	64.0	64.5	65.0	65.5	66.0	66.5	67.0	67.5
45.0										
45.5										
46.0										
46.5										
47.0										
47.5	19	19								
48.0	20	20	20	19						
48.5	21	21	20	20	20	19				
49.0	22	21	21	21	20	20	20	19	19	
49.5	22	22	22	21	21	21	20	20	20	19
50.0	23	23	22	22	22	21	21	21	21	20
50.5	24	23	23	23	23	22	22	22	21	21
51.0	25	24	24	24	23	23	23	22	22	22
51.5	25	25	25	24	24	24	23	23	23	22
52.0	26	26	25	25	25	24	24	24	23	23
52.5	27	26	26	26	25	25	25	24	24	24
53.0	27	27	27	26	26	26	25	25	25	24
53.5	28	28	27	27	27	26	26	26	25	25
54.0	29	28	28	28	27	27	27	26	26	26
54.5	29	29	29	28	28	28	27	27	27	26
55.0	30	30	29	29	29	28	28	28	27	27
55.5	31	30	30	30	29	29	29	28	28	28
56.0	31	31	30	30	30	30	29	29	29	28
56.5	32	31	31	31	30	30	30	29	29	29
57.0	32	32	32	31	31	31	30	30	30	29
57.5	33	33	32	32	32	31	31	31	30	30
58.0	34	33	33	33	32	32	32	31	31	31
58.5	34	34	34	33	33	33	32	32	32	31
59.0	35	35	34	34	34	33	33	33	32	32
59.5	35	35	35	34	34	34	33	33	33	33
60.0	36	36	35	35	35	34	34	34	33	33
60.5	37	36	36	36	35	35	35	34	34	34
61.0	37	37	37	36	36	36	35	35	35	34
61.5	38	37	37	37	36	36	36	36	35	35
62.0	38	38	38	37	37	37	36	36	36	35

* Circumference Value = waist circumference + hip circumference - neck circumference (in inches)

SECTION 2: PERCENT BODY FAT ESTIMATION

PERCENT BODY FAT ESTIMATION FOR WOMEN

Circumference Value*	Height (in)									
	63.0	63.5	64.0	64.5	65.0	65.5	66.0	66.5	67.0	67.5
62.5	39	39	38	38	38	37	37	37	36	36
63.0	40	39	39	39	38	38	38	37	37	37
63.5	40	40	39	39	39	38	38	38	37	37
64.0	41	40	40	40	39	39	39	38	38	38
64.5	41	41	41	40	40	40	39	39	39	38
65.0	42	41	41	41	40	40	40	39	39	39
65.5	42	42	42	41	41	41	40	40	40	39
66.0	43	42	42	42	41	41	41	41	40	40
66.5	43	43	43	42	42	42	41	41	41	40
67.0	44	44	43	43	43	42	42	42	41	41
67.5	44	44	44	43	43	43	42	42	42	41
68.0	45	45	44	44	44	43	43	43	42	42
68.5	45	45	45	44	44	44	43	43	43	43
69.0	46	46	45	45	45	44	44	44	43	43
69.5	46	46	46	45	45	45	44	44	44	44
70.0	47	47	46	46	46	45	45	45	44	44
70.5			47	46	46	46	46	45	45	45
71.0				47	47	46	46	46	45	45
71.5						47	47	46	46	46
72.0							47	47	46	46
72.5									47	47
73.0										
73.5										
74.0										
74.5										
75.0										
75.5										
76.0										
76.5										
77.0										
77.5										
78.0										
78.5										
79.0										
79.5										

* Circumference Value = waist circumference + hip circumference - neck circumference (in inches)

PERCENT BODY FAT ESTIMATION FOR WOMEN

Circumference Value*	Height (in)									
	68.0	68.5	69.0	69.5	70.0	70.5	71.0	71.5	72.0	72.5
45.0										
45.5										
46.0										
46.5										
47.0										
47.5										
48.0										
48.5										
49.0										
49.5	19									
50.0	20	20	19							
50.5	21	20	20	20	19	19				
51.0	21	21	21	20	20	20	19	19		
51.5	22	22	21	21	21	20	20	20	20	19
52.0	23	22	22	22	21	21	21	21	20	20
52.5	23	23	23	22	22	22	22	21	21	21
53.0	24	24	23	23	23	22	22	22	22	21
53.5	25	24	24	24	23	23	23	23	22	22
54.0	25	25	25	24	24	24	24	23	23	23
54.5	26	26	25	25	25	24	24	24	24	23
55.0	27	26	26	26	25	25	25	25	24	24
55.5	27	27	27	26	26	26	25	25	25	25
56.0	28	28	27	27	27	26	26	26	25	25
56.5	29	28	28	28	27	27	27	26	26	26
57.0	29	29	29	28	28	28	27	27	27	26
57.5	30	29	29	29	29	28	28	28	27	27
58.0	30	30	30	29	29	29	29	28	28	28
58.5	31	31	30	30	30	29	29	29	29	28
59.0	32	31	31	31	30	30	30	29	29	29
59.5	32	32	32	31	31	31	30	30	30	29
60.0	33	32	32	32	32	31	31	31	30	30
60.5	33	33	33	32	32	32	32	31	31	31
61.0	34	34	33	33	33	32	32	32	32	31
61.5	35	34	34	34	33	33	33	32	32	32
62.0	35	35	35	34	34	34	33	33	33	32

* Circumference Value = waist circumference + hip circumference - neck circumference (in inches)

PERCENT BODY FAT ESTIMATION FOR WOMEN

Circumference Value*	Height (in)									
	68.0	68.5	69.0	69.5	70.0	70.5	71.0	71.5	72.0	72.5
62.5	36	35	35	35	34	34	34	34	33	33
63.0	36	36	36	35	35	35	34	34	34	34
63.5	37	37	36	36	36	35	35	35	34	34
64.0	37	37	37	36	36	36	36	35	35	35
64.5	38	38	37	37	37	36	36	36	36	35
65.0	38	38	38	38	37	37	37	36	36	36
65.5	39	39	38	38	38	37	37	37	37	36
66.0	40	39	39	39	38	38	38	37	37	37
66.5	40	40	39	39	39	39	38	38	38	37
67.0	41	40	40	40	39	39	39	39	38	38
67.5	41	41	41	40	40	40	39	39	39	38
68.0	42	41	41	41	40	40	40	40	39	39
68.5	42	42	42	41	41	41	40	40	40	39
69.0	43	42	42	42	41	41	41	41	40	40
69.5	43	43	43	42	42	42	41	41	41	41
70.0	44	43	43	43	43	42	42	42	41	41
70.5	44	44	44	43	43	43	42	42	42	42
71.0	45	44	44	44	44	43	43	43	42	42
71.5	45	45	45	44	44	44	43	43	43	43
72.0	46	45	45	45	45	44	44	44	43	43
72.5	46	46	46	45	45	45	44	44	44	44
73.0	47	46	46	46	45	45	45	45	44	44
73.5		47	47	46	46	46	45	45	45	44
74.0				47	46	46	46	46	45	45
74.5					47	47	46	46	46	45
75.0							47	46	46	46
75.5								47	47	46
76.0										47
76.5										
77.0										
77.5										
78.0										
78.5										
79.0										
79.5										

* Circumference Value = waist circumference + hip circumference - neck circumference (in inches)

PERCENT BODY FAT ESTIMATION FOR WOMEN

Circumference Value*	Height (in)									
	73.0	73.5	74.0	74.5	75.0	75.5	76.0	76.5	77.0	77.5
45.0										
45.5										
46.0										
46.5										
47.0										
47.5										
48.0										
48.5										
49.0										
49.5										
50.0										
50.5										
51.0										
51.5										
52.0	20	19	19							
52.5	20	20	20	19	19					
53.0	21	21	20	20	20	20	19	19		
53.5	22	21	21	21	21	20	20	20	19	19
54.0	22	22	22	21	21	21	21	20	20	20
54.5	23	23	22	22	22	22	21	21	21	20
55.0	24	23	23	23	22	22	22	22	21	21
55.5	24	24	24	23	23	23	23	22	22	22
56.0	25	25	24	24	24	23	23	23	23	22
56.5	26	25	25	25	24	24	24	24	23	23
57.0	26	26	26	25	25	25	24	24	24	24
57.5	27	26	26	26	26	25	25	25	25	24
58.0	27	27	27	27	26	26	26	25	25	25
58.5	28	28	27	27	27	27	26	26	26	25
59.0	29	28	28	28	27	27	27	27	26	26
59.5	29	29	29	28	28	28	27	27	27	27
60.0	30	30	29	29	29	28	28	28	28	27
60.5	30	30	30	30	29	29	29	28	28	28
61.0	31	31	30	30	30	30	29	29	29	28
61.5	32	31	31	31	30	30	30	30	29	29
62.0	32	32	32	31	31	31	30	30	30	30

* Circumference Value = waist circumference + hip circumference - neck circumference (in inches)

PERCENT BODY FAT ESTIMATION FOR WOMEN

Circumference Value*	Height (in)									
	73.0	73.5	74.0	74.5	75.0	75.5	76.0	76.5	77.0	77.5
62.5	33	32	32	32	32	31	31	31	30	30
63.0	33	33	33	32	32	32	32	31	31	31
63.5	34	34	33	33	33	32	32	32	32	31
64.0	34	34	34	34	33	33	33	32	32	32
64.5	35	35	34	34	34	33	33	33	33	32
65.0	35	35	35	35	34	34	34	33	33	33
65.5	36	36	35	35	35	35	34	34	34	33
66.0	37	36	36	36	35	35	35	35	34	34
66.5	37	37	37	36	36	36	35	35	35	35
67.0	38	37	37	37	36	36	36	36	35	35
67.5	38	38	38	37	37	37	36	36	36	36
68.0	39	38	38	38	38	37	37	37	36	36
68.5	39	39	39	38	38	38	37	37	37	37
69.0	40	39	39	39	39	38	38	38	37	37
69.5	40	40	40	39	39	39	39	38	38	38
70.0	41	40	40	40	40	39	39	39	38	38
70.5	41	41	41	40	40	40	40	39	39	39
71.0	42	41	41	41	41	40	40	40	39	39
71.5	42	42	42	41	41	41	41	40	40	40
72.0	43	42	42	42	42	41	41	41	40	40
72.5	43	43	43	42	42	42	42	41	41	41
73.0	44	43	43	43	43	42	42	42	41	41
73.5	44	44	44	43	43	43	42	42	42	42
74.0	45	44	44	44	44	43	43	43	42	42
74.5	45	45	45	44	44	44	43	43	43	43
75.0	46	45	45	45	44	44	44	44	43	43
75.5	46	46	46	45	45	45	44	44	44	44
76.0	47	46	46	46	45	45	45	45	44	44
76.5		47	46	46	46	46	45	45	45	44
77.0			47	47	46	46	46	45	45	45
77.5					47	47	46	46	46	45
78.0						47	47	46	46	46
78.5								47	47	46
79.0										47
79.5										

* Circumference Value = waist circumference + hip circumference - neck circumference (in inches)

APPENDIX J

PHYSICAL FITNESS TEST (PFT) ENHANCEMENT TRAINING

1. Background. The PFT is an excellent instrument for measuring a Marine's general fitness level, as it relates to upper body strength and cardio-vascular endurance. Although the PFT is not specifically designed to evaluate a Marine's level of combat fitness, it does provide a baseline for potential success in meeting the physical demands of combat. This Appendix provides an abridged guide for PFT training that both reviews and applies the general principles of strength and conditioning for each of the PFT events.

2. General Strength and Conditioning Principles

a. Specificity. The main benefit of specificity is specific training develops neuromuscular efficiency. Movement that is more efficient means more repetitions or faster times. This principle states that to become good at a particular movement, you must practice that particular movement. For example, a pistol shooter does not become good at rapid fire by shooting continual precision strings of fire. Specificity states the shooter would be better served by specific training because specific training yields specific gains.

(1) Pullup/Flexed Arm Hang. There are many exercises and machines that can enhance upper body strength, but there is no better exercise for improving pullup or the flexed-arm hang performance than the act of doing or practicing them.

(2) Crunch. The various abdominal strengthening devices (machines) and routines that focus on the mid- to lower abdominal muscles will improve crunch performance in most cases. However, doing or practicing crunches is the best method for improving crunch performance.

(3) 3.0 Mile Run. Exercise machines such as stair climbers, stationary bicycles, elliptical cross trainers, and rowing machines can enhance a Marine's cardiovascular endurance. They cannot, however replace running as the activity that will best improve running performance.

b. Overload. The acronym "SAID" stands for "Specific Adaptations to Imposed Demands," and describes what happens when the muscular and cardiovascular systems are overloaded or "pushed." When a Marine does an activity to a point of fatigue, muscle systems that have been exerted will become stronger and better able to handle the increased exercise stimulus. Proper rest and nutrition are important cofactors in ensuring the benefits of overloading are realized. It is important for the Marine and the commander to understand that improper overloading could have a detrimental effect on the desired result.

(1) Pullup/Flexed-Arm Hang. Routines that require completing set(s) of pullups to the point of exhaustion are worthwhile. Form and slowed execution of the pullup should be maintained in order to facilitate momentary muscular fatigue. Executing pullups under weight (e.g., with a pack or flak jacket on) and varying the width of the grip (e.g., narrow to wide grip) are excellent training methods. Training for pullups in this manner will help ensure that a Marine will be able to do proper pullups in an all-out effort on test day.

(2) Crunch. Applying the principle of overload in preparing for the crunch requires doing two sets of crunches until momentary muscular fatigue. Tracking the amount of "complete" crunches a Marine can do until the point of fatigue, in a certain time period, will provide excellent feedback on the effectiveness of the overload principle. A second way to train for the crunch event would be for a Marine to do as many crunches as he can, without stopping, until complete muscle failure. Focusing on using the abdominal muscles (keeping them tight throughout the exercise), and executing them in slow motion is an excellent way to intensify the training. Track the number of proper repetitions executed to determine effectiveness.

(3) 3.0 Mile Run. Interval training done once or twice a week can enhance a Marine's running performance. Interval training sessions take less time, but require a higher level of intensity to run the shorter distances. The 400-meter and 800-meter courses are the best distances for increasing performance for the 3.0 mile run. If the distance interval training were longer than 800 meters, the Marine would not be able to run at the targeted speed. If the distance is shorter than 400 meters, the Marine cannot reach the targeted level of intensity needed for endurance training.

(a) To calculate the time an individual Marine should run the 400/800-meter distances during an interval training session is as follows:

1 Determine the most recent 3.0 mile run time;

2 Divide that time by 3. This gives the mile-split time for the run;

3 Divide the mile-split time by 2. Subtract that time by 10-20 seconds. This gives the time the 800-meter intervals should be run.

4 To determine the 400-meter interval time, divide the mile-split time by 4. Then subtract 5-10 seconds. This is the time the 400-meter intervals should be run.

(b) Calculating rest in between each running exercise is very important. Complete recovery in between the intervals lessens the benefit of overload, but not enough rest is also detrimental. Incomplete recovery in between intervals should be the goal. The work/rest ratio for 800-meter intervals should be 1:1. In other words, active rest in the form of light jogging or walking (not stopping) should be equal to the time that was run. For example if a 4:00 minute interval time was run, then rest should be 4:00 minutes. The work/rest ratio for 400-meter intervals should be 1:2. Using the same type of rest as in 800-meter interval session, a 1 min 40 second 400-meter interval would be followed by a 3 min 20 second rest period.

1 Example: A Marine wants to do interval sessions to improve the 22:00 he ran in the last PFT.

a Divide 22:00 by 3 (7:20);

b Divide 7:20 by 2 to determine the pace for an 800-meter split time for a 7:20 mile pace. (3:40);

c Subtract 10-20 seconds to determine the pace for 800-meter intervals. (3:20-3:40);

d The rest between these 800 repeats should be 3:20-3:40.

e Determine the pace for a 400-meter interval by dividing 7:20 by 4 to determine the pace for a 400-meter split time for 7:20 mile pace (1:50). Subtract 5-10 seconds to determine pace for 400-meter intervals. (1:40-1:45);

f The rest between these 400 repeats should be 3:20-3:30.

2 A sample (beginner's) interval workout:

- Daily 16 Warmup
- 800 meter jog
- 800-rest-800-rest-400-rest-400
- 800 meter jog
- Daily 16 Cool-Down

(c) The total mileage for this workout is 2.5 miles with 1.5 miles being speed-work. Many Marines and recreational runners do intervals that are much too fast for their goal time. Keeping the predetermined goal pace should be the goal for this workout. A Marine should not feel totally exhausted following this training session. As a Marine gets more comfortable with this training, more intervals can be added.

(d) Interval training should be combined with other types of running, during the weekly training schedule, for the training to be effective. Longer distance runs along with tempo running (faster running but not at race pace), combined with interval running will improve most Marines' 3.0 mile run times.

3. Summary

a. Specificity and overload are two of the most important strength and conditioning principles. Other principles include:

(1) Regularity - Strength and cardiovascular training should be done 3-4 times a week.

(2) Variety - Varying training regimes periodically prevents boredom.

(3) Recovery - Rest is physiologically necessary for growth and repair.

(4) Balance - A good training plan should exercise all major muscle groups, and include warmup and cool-down phases.

(5) Progression - Ensures that systems are continually challenged.

b. There are many ways in which Marines can improve their performance on the PFT. By applying the strength and conditioning principles listed above, training can be enhanced to improve not only PFT performance but increase a Marines' physical readiness for combat.

APPENDIX K

LIGHT/LIMITED DUTY ASSIGNMENT PROCESS

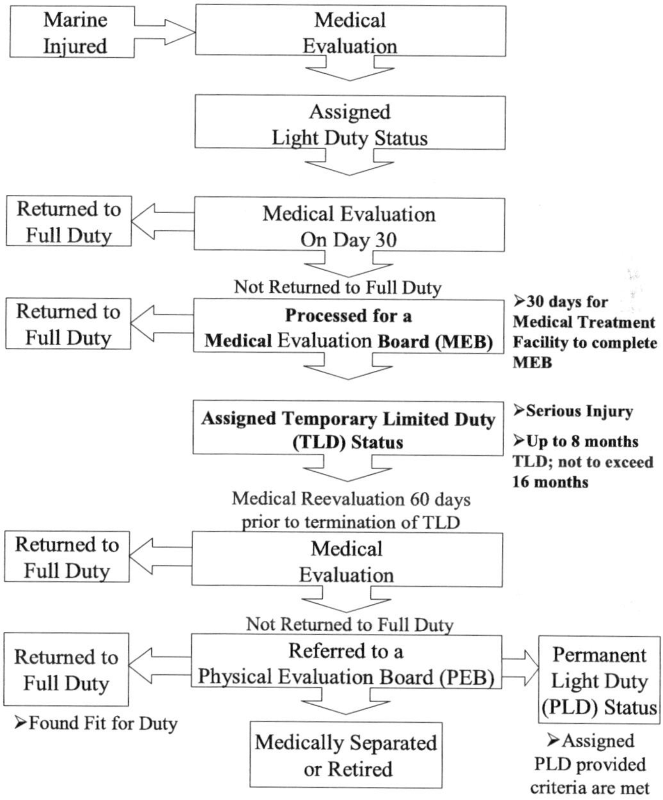

NOTE:

- Marines on light or limited duty (TLD & PLD) are required to maintain Height/Weight or Body Fat Standards regardless of duty status. Marines not maintaining standards can be assigned to the Body Composition Program.
- Marines assigned to light or limited duty (TLD & PLD) who are medically restricted from completing all three events of a PFT will be screened for participation in a Partial PFT – 2 of the 3 events.

APPENDIX L

PREGNANCY PFT AND BODY COMPOSITION PROGRAM FLOWCHART

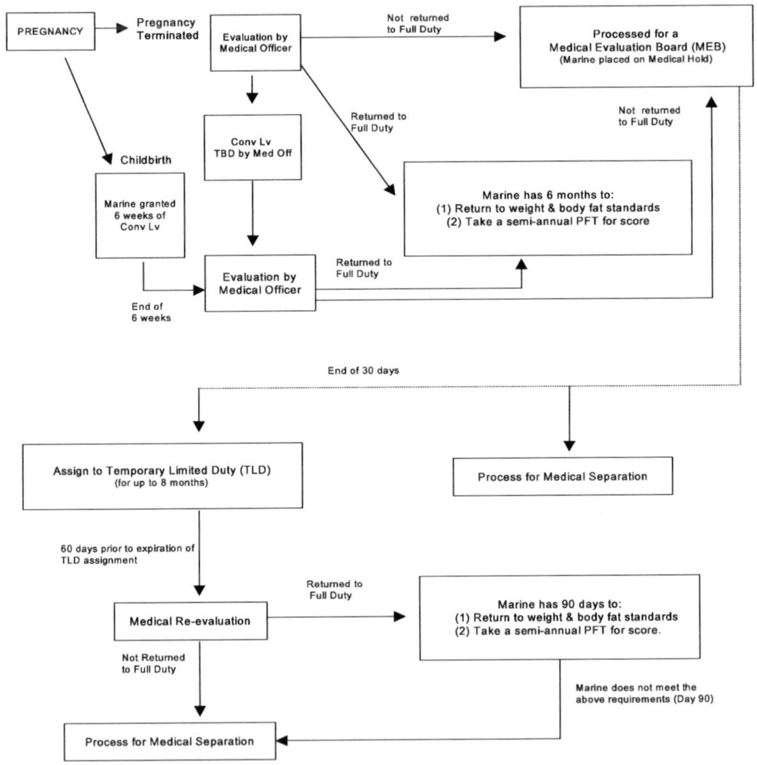

NOTE: Marines assigned to a Body Composition Program (BCP) PRIOR to pregnancy will go into an "inactive status". Once returned to full duty, Marine will return to an active status and will have 6 months from that day to return to weight & body fat standards, regardless on time spent on Program previously.

Headquarters
Department of the Army
Washington, DC
10 June 1987

Army Regulation 600–9

Effective 10 June 1987

Personnel—General

The Army Weight Control Program

By Order of the Secretary of the Army:

JOHN A. WICKHAM, JR.
General, United States Army
Chief of Staff

Official:

R. L. DILWORTH
Brigadier General, United States Army
The Adjutant General

History. The original form of this regulation was published on 1 September 1986. Since that time, changes have been issued to amend the original. As of 3 August 1987, permanent changes 1 and 2 remained in effect. This UPDATE printing incorporates all of those changes directly into the body of text. This publication has been reorganized to make it compatible with the Army electronic publishing database. No content has been changed.

Summary. This regulation implements guidance in DOD directive 1308.1, dated 29 June 1981, which establishes a weight control program in all the Services.

Applicability. This regulation applies to all members of the Active Army, the Army National Guard (ARNG), and the U.S. Army Reserve (USAR) to include those ARNG and USAR personnel in Active Guard/Reserve (AGR) status.

Proponent and exception authority. The proponent agency of this regulation is the Office of the Deputy Chief of Staff for Personnel.

Army management control process. This regulation is subject to the requirements of AR 11–2. It contains internal control provisions but does not contain checklists for conducting internal control reviews. These checklists will be contained in a DA Circular in the 11 series.

Supplementation. Supplementation of this regulation is prohibited without prior approval from HQDA (DAPE–MPA), WASH DC 20310.

Interim changes. Interim changes to this regulation are not official unless they are authenticated by The Adjutant General. Users will destroy interim changes on their expiration dates unless sooner superseded or rescinded.

Suggested Improvements. Users are invited to send comments and suggested improvements on DA Form 2028 (Recommended Changes to Publications and Blank Forms) directly to HQDA (DAPE–MPA), WASH DC 20310.

Distribution. Active Army, ARNG, and USAR: A.

Contents (Listed by paragraph and page number)

Section I
Introduction, page 1
Purpose • 1, *page 1*
References • 2, *page 1*
Explanation of abbreviations and terms • 3, *page 1*
Objectives • 4, *page 1*

Section II
Responsibilities, page 1
General • 5, *page 1*
Deputy Chief of Staff for Personnel (DCSPER) • 6, *page 1*
The Surgeon General (TSG) • 7, *page 1*
Deputy Chief of Staff for Logistics (DCSLOG) • 8, *page 1*
Chief, National Guard Bureau (CNGB) • 9, *page 1*
Chief, Army Reserve (CAR) • 10, *page 2*
Commanders of major Army commands (MACOMs) • 11, *page 2*
Commanding General, US Army Training and Doctrine Command (CG, TRADOC) • 12, *page 2*

UNCLASSIFIED

Contents—Continued

Commanding General, US Army Forces Command (CG, FORSCOM) • 13, *page 2*
Commanders of major medical commands • 14, *page 2*
Commanding General, US Army Reserve Components Personnel Center (CG, ARPERCEN) • 15, *page 2*
Individuals • 16, *page 2*
Commanders and superviors • 17, *page 2*
Health Care Personnel • 18, *page 3*
Master Fitness Trainers • 19, *page 3*

Section III
Proper Weight Control, page 3
Policy • 20, *page 3*
Procedures • 21, *page 4*
Reenlistment criteria • 22, *page 6*

Appendixes

A. References, *page 11*

B. Standard Methods for Determining Body Fat Using Body Circumferences, Height and Weight, *page 12*

C. Nutrition Guide to the Weight Control Program, *page 25*

Table List

Table 2: Maximum allowable percent body fat standards, *page 3*
Table 1: Weight for Height Table (Screening Table Weight), *page 7*
Table B–1: Male Abdomen and Neck Factor, *page 15*
Table B–2: Male Height Factor, *page 16*
Table B–3: Female Weight Factor, *page 17*
Table B–4: Female Height Factor, *page 17*
Table B–5: Female Hip Factor, *page 18*
Table B–6: Female Forearm Factor, *page 19*
Table B–7: Female Neck Factor, *page 19*
Table B–8: Female Wrist Factor, *page 19*
Table C–0–1: Nutrition, *page 25*
Table C–0–2: Diet structures, *page 26*
Table C–0–3: Weight maintenance formula, *page 26*
Table C–0–4: Weight loss formula, *page 27*
Table C–0–5: Nutrient-calories guide, *page 27*
Table C–0–6: Vitamins–Minerals–Water, *page 27*
Table C–0–7: High calorie-low nutrient foods, *page 28*
Table C–0–8: Equal measurements, *page 28*
Table C–0–9: Alcohol calories, *page 29*
Table C–0–10: Sample calorie chart, *page 30*
Table C–1: Weight Control a Personal Choice, *page 31*
Table C–2: Comparison of Regular and TRIMM* meals, *page 31*
Table C–3: Sample 1200 Calorie Menu, *page 32*
Table C–4: Sample 1500—Calorie Menu, *page 32*

Figure List

Figure 1: Sample correspondence for Weight Control Program, *page 8*
Figure 1: Sample correspondence for Weight Control Program—Continued, *page 9*
Figure 2: Flow process guide for screening and weight control actions (applies to all personnel; officers, warrant officers, enlisted), *page 10*
Figure B–1: Sample of a completed DA Form 5500, *page 20*
Figure B–2: Male Measurement, *page 21*
Figure B–2: Male Measurement—Continued, *page 22*

Contents—Continued

Figure B–3: Sample of a completed DA Form 5501, *page 23*
Figure B–4: Female Measurement, *page 24*

Glossary

Section I
Introduction

1. Purpose
This regulation establishes policies and procedures for the implementation of the Army Weight Control Program.

2. References
Required publications are listed in appendix A.

3. Explanation of abbreviations and terms
Abbreviations and special terms used in this regulation are explained in the glossary.

4. Objectives
a. The primary objective of the Army Weight Control Program is to insure that all personnel—
(1) Are able to meet the physical demands of their duties under combat conditions.
(2) Present a trim military appearance at all times.
b. Excessive body fat—
(1) Connotes a lack of personal discipline.
(2) Detracts from military appearance.
(3) May indicate a poor state of health, physical fitness, or stamina.
c. Objectives of the Army Weight Control Program are to—
(1) Assist in establishing and maintaining—
(a) Discipline.
(b) Operational readiness.
(c) Optimal physical fitness.
(d) Health.
(e) Effectiveness of Army personnel through proper weight control.
(2) Establish appropriate body fat standards.
(3) Provide procedures for which personnel are counseled to assist in meeting the standards prescribed in this regulation.
(4) Foster high standards of professional military appearance expected of all personnel.

Section II
Responsibilities

5. General
The Army traditionally has fostered a military appearance which is neat and trim. Further, an essential function of day-to-day effectiveness and combat readiness of the Army is that all personnel are healthy and physically fit. Self-discipline to maintain proper weight distribution and high standards of appearance are essential to every individual in the Army.

6. Deputy Chief of Staff for Personnel (DCSPER)
The DCSPER is responsible for the Army Weight Control Program.

7. The Surgeon General (TSG)
TSG will—
a. Establish medical examination and counseling policies in support of the Army Weight Control Program.
b. Evaluate the medical aspects of the program.
c. Establish and review procedures for determination of body fat content.
d. Provide guidance on improving the nutritional status of soldiers.

8. Deputy Chief of Staff for Logistics (DCSLOG)
The DCSLOG will—
a. Establish food service guidance in support of the Army Weight Control Program.
b. Publish guidance and information pertaining to the caloric content of items served on master menus.

9. Chief, National Guard Bureau (CNGB)
The CNGB will—
a. Implement and monitor the Army Weight Control Program in the ARNG (including units).

b. Take appropriate action under guidance prescribed in this regulation.

10. Chief, Army Reserve (CAR)

The CAR will—

a. Monitor the Army Weight Control Program in the USAR.

b. Take appropriate action under guidance prescribed in this regulation.

11. Commanders of major Army commands (MACOMs)

MACOM commanders will insure that soldiers within their commands are evaluated under the body fat standards prescribed in this regulation.

12. Commanding General, US Army Training and Doctrine Command (CG, TRADOC)

The CG, TRADOC will inform personnel at initial entry on active duty (AD) about basic nutrition and sound food consumption practices.

13. Commanding General, US Army Forces Command (CG, FORSCOM)

The CG, FORSCOM will implement and maintain the Army Weight Control Program in USAR troop program units (TPUs).

14. Commanders of major medical commands

a. Commanding General, US Army Health Services Command (CG, HSC). The CG, HSC will—

(1) Institute weight reduction and counseling programs in Army medical facilities in support of the Army Weight Control Program.

(2) Provide appropriate literature and training aids for use by soldiers, supervisors, and commanders in selection of a proper diet.

b. Commanders of other major medical commands (overseas) will institute weight reduction and counseling programs in Army medical facilities in support of the Army Weight Control Program overseas.

15. Commanding General, US Army Reserve Components Personnel Center (CG, ARPERCEN)

The CG, ARPERCEN will—

a. Monitor the Army Weight Control Program in the Individual Ready Reserve (IRR).

b. Take appropriate action under guidance prescribed in this regulation.

c. Insure that members applying for tours of AD, active duty for training (ADT), active duty support (ADS), and AGR meet the body fat standards prescribed in this regulation. (Soldiers who do not meet these standards will not be permitted to enter on AD, ADT, or ADS, or in AGR status.)

16. Individuals

Each soldier (Commissioned Officer, Warrant Officer and Enlisted) is responsible for meeting the standards presecribed in this regulation. To assist soldiers in meeting these responsibilities, screening tables will continue to be prescribed for use as is currently being done. A five percent zone below the screening table weight ceiling is suggested as a help to soldiers in targeting their personal weight at a level which will minimize the probability of exceeding the screening table weight ceiling as a matter of habit. Soldiers should be coached to select their personal weight goal within or below the five percent zone and strive to maintain that weight through adjustment of life style and fitness routines. If a soldier consistently exceeds the personal weight goal, he or she should seek the assistance of master fitness trainers for advice in proper exercise and fitness; and health care personnel for a proper dietary program. In other words, exceeding a properly selected goal should "trigger" the soldier to use the substantial help available to alter the fitness and dietary behavior before confronting the finality of the screening table and initiation of official action if the body fat standards are exceeded.

17. Commanders and supervisors

Commanders and supervisors (Active and Reserve Components) will—

a. Implement the Army Weight Control Program, to include evaluation of the weight and military appearance of all soldiers under their jurisdiction; to include measuring body fat as prescribed in this regulation.

b. Insure the continued evaluation of all soldiers under their command or supervision against the body fat standards prescribed in this regulation.

c. Maintain data on soldiers in their command or under their supervision who—

(1) Enter a weight control program each year.

(2) Subsequently either meet the body fat standards prescribed in this regulation or were separated from the Service for reasons related to overweight conditions.

d. Encourage soldiers to establish a personal weight goal as described above.

e. ensure that personnel responsible for issuing TDY and PCS orders include in the text of all orders the words: "You are responsible for reporting to your next duty station/school in satisfactory physical condition, able to pass the APFT and meet weight standards."

18. Health Care Personnel

Health care personnel will—

a. Assist commanders and supervisors by providing weight reduction counseling to individuals who are overweight.

b. Identify those individuals who have a pathological condition requiring medical treatment.

c. Evaluate overweight soldiers, when—

(1) A soldier has a medical limitation (DA Form 3349, AR 40–501 prescribes assignment limitations for soldiers with profiles. For example, no mandatory strenuous physical activity); or

(2) Is pregnant; or

(3) When an evaluation is requested by a unit commander (this is an option for unit commanders and is not mandatory); or

(4) Separation is being considered for failure to make satisfactory progress in a weight control program; or

(5) Six months prior to ETS.

19. Master Fitness Trainers

Master Fitness Trainers will—

a. Prescribe proper exercise and fitness techniques to assist soldiers in determining, achieving and maintaining an appropriate personal weight goal.

b. Assist commanders in developing proactive programs which clearly establish physical fitness as a unit value.

Section III
Proper Weight Control

20. Policy

a. Commanders and supervisors will monitor all members of their command (officers, warrant officers, and enlisted personnel) to insure that they maintain proper weight, body composition (as explained in the glossary), and personal appearance. At minimum, personnel will be weighed when they take the Army Physical Fitness Test (APFT) or at least every 6 months. Personnel exceeding the screening table weight (shown in table 1) or identified by the commander or supervisor for a special evaluation will have a determination made of percent body fat. Identification and counseling of overweight personnel are required.

b. Commanders and supervisors will provide educational and other motivational programs to encourage personnel to attain and maintain proper weight (body fat) standards. Such programs will include—

(1) Nutrition education sessions conducted by qualified health care personnel. These sessions are required for all soldiers enrolled in a Weight Control Program.

(2) Exercise programs, even though minimum APFT standards are achieved.

c. Maximum allowable percent body fat standards are as follows in table 2. However, all personnel are encouraged to achieve the more stringent Department of Defense (DOD)-wide goal, which is 20 percent body fat for males and 26 percent body fat for females.

Table 2
Maximum allowable percent body fat standards

Age Group	Male (% body fat)	Female (% body fat)
17–20	%20	%28
21–27	%22	%30
28–39	%24	%32
40 & Older	%26	%34

d. Personnel who are overweight (as explained in the glossary) (including soldiers who become pregnant while on the weight control program)—

(1) Will be considered nonpromotable (to the extent such nonpromotion is permitted by law).

(2) Will not be authorized to attend professional military or civilian schooling. Personnel arriving at professional military schools overweight will be denied enrollment, except:

(a) for personnel in a TDY and return or a TDY enroute to PCS status who, in the opinion of the appropriate school commander or commandant, can potentially meet their body fat standard by the end of the course or within 30 days, whichever comes first. Overweight personnel who do not meet their body fat standard within the probationary period will be disenrolled and will not graduate. An academic efficiency report as required, will be annotated to indicate failure to meet body fat standards, and the appropriate career division will be notified for those personnel in a TDY enroute to PCS status.

(b) for personnel in a PCS status, and then, only for a 30 day probationary period. If at the end of the probationary period, body fat standards have not been met, the student will be disenrolled. An academic efficiency report as required, will be annotated to indicate failure to meet body fat standards, and the appropriate career division will be notified for assignment instructions.

(3) Will not be assigned to command positions.

21. Procedures

a. Body fat composition will be determined for personnel—

(1) Whose body weight exceeds the screening table weight in table 1.

(2) When the unit commander or supervisor determines that the individual's appearance suggests that body fat is excessive.

b. Routine weigh-ins will be accomplished at the unit level. Percent body fat measurements will be accomplished by company or similar level commanders (or their designee) in accordance with standard methods prescribed in Appendix B to this regulation. Soldiers will be measured by individuals of the same gender. If this cannot be accomplished, a female soldier will be present when males measure females. IRR members on AT, ADT, and SADT will have a weigh-in, and body fat evaluation (if required) by the unit to which attached. Active and Reserve component soldiers exceeding the body fat standards in paragraph 20*c*, above, will be provided weight reduction counseling by health care personnel, entered in a Weight Control Program by unit commanders, and flagged under the provisions of AR 600–31 by the unit commander. Enrollment in a Weight Control Program starts on the day that the soldier is informed by the unit commander that he/she has been entered in a weight control program. The weight reduction counseling can be accomplished prior to or shortly after entry into a program.

c. The sample correspondence shown in figure 1 will be completed and retained by the unit commander or supervisor to document properly recommendations and actions taken in each case.

d. A medical evaluation will be accomplished by health care personnel when the soldier has a medical limitation, or is pregnant, or when requested by the unit commander. One is also required for soldiers being considered for separation due to failure to make satisfactory progress in a weight control program, or within 6 months of ETS. Aircraft crew members, who exceed the body fat standards, will be referred to a flight surgeon for possible impact on flight status. If an individual's condition is diagnosed by medical authorities to result from an underlying or associated disease process, health care personnel will take one of the following actions:

(1) Prescribe treatment to alleviate the condition and return personnel to their unit.

(2) Hospitalize individuals for necessary treatment; this action applies to Active Army personnel only. Reserve Component personnel will be referred to their personal physician for further evaluation or treatment at the individual's expense.

(3) Determine whether the individual's condition is medically disqualifying for continued service. In these cases, disposition will be made under provisions of appropriate regulations.

e. If health care personnel discover no underlying or associated disease process as the cause of the condition (or RC personnel decline to be evaluated under paragraph 21*d*) and the individual is classified as overweight, these facts will be documented and the individual entered in a weight control program by the unit commander. Suspension of favorable personnel actions will be initiated under AR 600–31 for personnel in a weight control program.

(1) The required weight loss goal of 3 to 8 pounds per month is considered a safely attainable goal to enable soldiers to lose excess body fat and meet the body fat standards described in para 20*c*. Weight-ins will be made by unit personnel monthly (or during unit assemblies for ARNG and USAR personnel) to measure progress. A body fat evaluation may also be done by unit personnel to assist in measuring progress.

(2) As an exception to *g* below, an individual who has not made satisfactory progress after any two consecutive monthly weigh-ins may be referred by the commander or supervisor to health care personnel for evaluation or reevaluation. If health care personnel are unable to determine a medical reason for lack of weight loss—and if the individual is not in compliance with the body fat standards at paragraph 20*c* and still exceeds the screening table weight (table 1)—the commander or supervisor will inform the individual that—

(a) Progress is unsatisfactory.

(b) He or she is subject to separation as specified in *j* below.

f. Commanders and supervisors will remove individuals administratively from a weight control program as soon as the body fat standard is achieved. The screening table weight will not be used to remove soldiers from a Weight Control Program. The removal action will be documented as shown in figure 1; removal of suspension of favorable personnel actions will be accomplished at that time.

g. After a period of dieting and/or exercise for 6 months, soldiers who have not made satisfactory progress (as explained in the glossary) and who still exceed the screening table and body fat standards will be processed as follows:

(1) If health care personnel determine that the condition is due to an underlying or associated disease process, action described in *d* above will be taken.

(2) If no underlying or associated disease process is found to cause the overweight condition, the individual will be subject to separation from the Service under appropriate regulations indicated in *j* below.

h. Personnel will be continued in a weight control program (as provided in *e* through *g* above) after the initial 6-month period if they—

(1) Still exceed the body fat standard.

(2) Have made satisfactory progress toward their weight loss (as indicated in the glossary), or are at or below the screening table weight (table 1).

(3) *For RC personnel only.* If the individual has not obtained an evaluation from his/her personal physician under the provisions of 21*d* above, and cannot demonstrate that the overweight condition results from an underlying or associated disease process, the individual may be separated under appropriate regulations without further medical evaluation by health care personnel.

i. To assist commanders and supervisors, a flow chart outlining procedural guidance is shown at figure 2.

j. The commander or supervisor will inform the individual in writing that initiation of separation proceedings is being considered under the following regulations: AR 635–200, chapter 5–15, AR 635–100, chapter 5; NGR 600–200, chapter 7; NGR 600–101; NGR 600–5; NGR 635–100; AR 135–175; or AR 135–178. This procedure will be followed unless a medical reason is found to preclude the loss of weight or there is other good cause to justify additional time in the weight control program.

(1) The individual will immediately respond to the separation consideration letter in writing. The commander or supervisor will consider the response and initiate separation action if no adequate explanation is provided, unless the individual submits an application for retirement, if eligible. USAR personnelwho fall under the purview of this paragraph will be released from AD or discharged, as appropriate.

(2) If separation action is not initiated or does not result in separation, the individual will be entered or continued in a weight control program, as specified in *e* above.

k. Following removal from a weight control program, if it is determined (under *a* above) that an individual exceeds the screening table weight (table 1) and the body fat standard prescribed in paragraph 20*c* within 36 months, the following will apply:

(1) If the unit commander determines that the individual exceeds the screening table weight and the body fat standard—

(a) Within 12 months from the date of the previous removal from the program and no underlying or associated disease process is found as the cause of the condition, the individual will be subject to separation from the Service under *j* above.

(b) After the 12th month, but within 36 months from the date of the previous removal from the program, and no underlying or associated disease process is found as the cause of the condition, the individual will be allowed 90 days to meet the standards. Personnel who meet the body fat standard after that period will be removed from the program. All others will be subject to separation from the Service under *j* above.

(c) Personnel who meet the AR 600–9 standards and become pregnant will be exempt from the standards for the duration of the pregnancy plus the period of convalescent leave after termination of pregnancy. They will be entered in a weight control program, if required, after completion of convalescent leave and approval of a medical doctor that they are fit for participation in a weight control program. This procedure also applies to individuals in a medical holding unit who have been hospitalized for long periods. Soldiers entered/reentered in a weight control program after pregnancy, prolonged treatment, or hospitalization will be considered to be in a new weight control program. Paragraph 21*k* of this regulation will not apply at that time.

(2) If the individual is determined to exceed the body fat standard and the condition is due to an underlying or associated disease process, action described in *d* above will be taken.

l. Inherent in the responsibility of selection boards is the obligation to select only those individuals who are considered to be physically fit to perform the duties required of them at all times. Compliance with the Army Weight Control Program as prescribed in this regulation will be considered in the selection process for promotion, professional military or civilian schooling, or assignment to command positions. Procedures for commanders and supervisors to provide current information for use by selection boards indicating whether individuals meet the prescribed standards will be included in DA regulations or issued by separate correspondence.

m. Records will be maintained in the MPRJ for personnel in weight control programs. On transfer from one unit to another, the losing commander or supervisor will forward a statement to the gaining unit with information indicating the status of the individual's participation in a weight control program. When the transfer is a permanent change of station, the unit commander's statement will be filed as a transfer documents in the Military Personnel Records Jacket, U.S. Army, under AR 640–10.

n. Upon removal from the weight control program, unit records on participation in a weight control program will be maintained in the MPRJ for a period of 36 months from date of removal. If the soldier is transferred to another unit prior to completion of 36 months, action will be taken in accordance with paragraph 21*m* above.

22. Reenlistment criteria

a. Personnel who exceed the screening table weight at table 1 and the body fat standard for their current age group in paragraph 20*c* will not be allowed to reenlist or extend their enlistment.

b. Exceptions to policy for Active Army personnel (including RC personnel on AD) are prescribed in this subparagraph. For soldiers who are otherwise physically fit and have performed their duties in a satisfactory manner, the commander exercising General Court Martial Convening Authority or the first general officer in the soldier's normal chain of command (whichever is in the most direct line to the soldier) may approve the following exceptions to policy:

(1) Extension of enlistment may be authorized for personnel who meet one of the following criteria:

(a) Individuals who have a temporary medical condition which precludes loss of weight. In such cases, the nature of on-going treatment will be documented; the extension will be for the minimum time necessary to correct the condition and achieve the required weight loss.

(b) Pregnant soldiers (except those soldiers who have medical conditions as listed in para 21*d*) who are otherwise fully qualified for reenlistment, including those with approved waivers, but who exceed acceptable standards prescribed in this regulation, will be extended for the minimum period which will allow birth of the child, plus 6 months. A clearance from the doctor tht the soldier is medically fit to participate in a Weight Control Program is required. Authority, which will be cited on DA Form 1695 (Oath of Extension of Enlistment), is AR 601–280, paragraph 3–3. On completion of the period of extension, the soldier will be reevaluated under paragraph 21.

(2) Exceptions to policy allowing reenlistment/extension of enlistment are authorized only in cases where—

(a) Medically documented conditions (para 21*d*) preclude attainment of required standards and

(b) Disability separation is not appropriate.

c. All requests for extension of enlistment for ARNG and USAR (TPU and IRR) personnel not on AD will be processed under NGR 600–200 or AR 140–111, chapter 3, as appropriate.

d. Requests for exceptions to policy will be forwarded through the chain of command, with the commander's personal recommendation and appropriate comment at each level. As a minimum, requests will include—

(1) The physician's evaluation.

(2) A record of progress in the weight control program.

(3) Current height and weight.

(4) Body fat content.

(5) Years of active Federal service.

(6) Other pertinent information.

e. Soldiers, who have completed a minimum of 18 years of Active Federal Service (AFS), may, if otherwise eligible, be extended for the minimum time required to complete 20 years AFS. Retirement must be accomplished not later than the last day of the month in which the soldier attains retirement eligibility. Application for retirement will be submitted at the time extension is authorized. Approval/disapproval authority is outlined in AR 601–280.

f. USAR soldiers, who have completed a minimum of 18 years of qualifying service for retired pay at age 60, may be extended for the minimum time required to complete 20 years qualifying service. Approval/disapproval authority is outlined in AR 140–111. Transfer to the IRR, Retired Reserve, or discharge will be accomplished at the end of the retirement year (RYE) in which the soldier attains the 20 qualifying years.

g. ARNG soldiers, who have completed a minimum of 18 years qualifying service for retired pay at age 60, may be extended for the minimum time required to complete 20 years qualifying service by the State TAG, disapproval authority is C, NGB. Transfer to the IRR, Retired Reserve, or discharge will be accomplished at the end of the retired year (RYE) in which the soldier attains the 20 qualifying years.

Table 1
Weight for Height Table (Screening Table Weight)

Height (in inches)	Male Age				Female Age			
	17–20	21–27	28–39	40+	17–20	21–27	28–39	40+
58	—	—	—	—	109	112	115	119
59	—	—	—	—	113	116	119	123
60	132	136	139	141	116	120	123	127
61	136	140	144	146	120	124	127	131
62	141	144	148	150	125	129	132	137
63	145	149	153	155	129	133	137	141
64	150	154	158	160	133	137	141	145
65	155	159	163	165	137	141	145	149
66	160	163	168	170	141	146	150	154
67	165	169	174	176	145	149	154	159
68	170	174	179	181	150	154	159	164
69	175	179	184	186	154	158	163	168
70	180	185	189	192	159	163	168	173
71	185	189	194	197	163	167	172	177
72	190	195	200	203	167	172	177	183
73	195	200	205	208	172	177	182	188
74	201	206	211	214	178	183	189	194
75	206	212	217	220	183	188	194	200
76	212	217	223	226	189	194	200	206
77	218	223	229	232	193	199	205	211
78	223	229	235	238	198	204	210	216
79	229	235	241	244	203	209	215	222
80	234	240	247	250	208	214	220	227

Notes:

1. The height will be measured in stocking feet (without shoes), standing on a flat surface with the chin parallel to the floor. The body should be straight but not rigid, similar to the position of attention. The measurement will be rounded to the nearest inch with the following guidelines:
   a. If the height fraction is less than 1/2 inch, round down to the nearest whole number in inches.
   b. If the height fraction is 1/2 inch or greater, round up to the next highest whole number in inches.
2. The weight should be measured and recorded to the nearest pound within the following guidelines.
   a. If the weight fraction is less than 1/2 pound, round down to the nearest pound.
   b. If the weight fraction is 1/2 pound or greater, round up to the next highest pound.
3. All measurements will be in a standard PT uniform (gym shorts and T-shirt, without shoes).
4. If the circumstances preclude weighing soldiers during the APFT, they should be weighed within 30 days of the APFT.
5. Add 6 pounds per inch for males over 80 inches and 5 pounds for females for each inch over 80 inches.

(Letterhead)

Office Symbol *Date*

SUBJECT: Weight Control Program

(Individual Soldier)

1. You have been determined to exceed the body fat standard and a goal of 3-8 pounds of weight loss per month is considered to be satisfactory progress. Failure to make satisfactory progress or achieve the body fat standards could result in separation from the service.

2. You have been flagged under the provisions of AR 600–31 and entered in a Weight Control Program.

Signature of Unit Commander

Office Symbol 1st End *Date*

FROM: Individual Soldier

TO: Commanding Officer

I understand my responsibilities to achieve the body fat standards to have my weight recorded periodically or during unit training assemblies (UTA) as applicable.

Soldier's Signature

Office Symbol 2d End

SUBJECT: Weight Control Program

FROM: Unit Commander

TO: MEDDAC

1. . . . exceeds the weight for height tables by . . . pounds and exceeds the body fat standards by .percent.

2. It is requested that a medical evaluation be conducted in view of the following (check applicable block):

( )Soldier's profile. ( )Pregnancy. ( )Unit Commander's special request. ( )Initiation of separation action (failure to make satisfactory progress in a Weight Control Program). ( )Within 6 months of ETS.

Signature of Unit Commander

Office Symbol 3d End

SUBJECT: Weight Control Program

FROM: Health Care Personnel

TO: Commanding Officer

1. In accordance with AR 600–9 . . . has been examined and found to be fit for participation in a Weight Control/Physical Exercise Program.

2. The cause of the overweight condition (is) (is not) due to a medical condition.

3. The following action(s) is/are recommended:

( )Initiation or continuation in a weight reduction program.

( )Medical treatment for pathological medical disorder (refer Reserve Component soldiers to their personal physician for evaluation at their own expense).

Signature of Health Care Personnel

Figure 1. Sample correspondence for Weight Control Program

Office Symbol 4th End

FROM: Unit Commander

TO: MEDDAC (Health Care Personnel)

1. . . . exceeds the screening table weight by . . . and body fat standards by . . . percent.

2. Nutrition education and weight reduction counseling are requested in accordance with AR 600–9 (para 20b (1)).

Signature of Unit Commander

Office Symbol 5th End

SUBJECT: Weight Control Program

FROM: Health Care Personnel

TO: Commanding Officer

1. . . . has been provided nutrition and weight reduction counseling in accordance with AR 600–9.

2. Follow-up counseling should be provided at unit level using information in Appendix C of this regulation and the assistance of master fitness trainers, if available.

Signature of Health Care Personnel

Office Symbol 6th End

FROM: Unit Commander

TO: Military Personnel Officer

1. The soldier has been determined to be in compliance with the provisions of AR 600–9, and is therefore removed from the Weight Control Program effective this date.

2. The individual's current weight is . . . pounds. Screening table weight ceiling is . . . pounds for present age category. Body fat content is . . . percent, which is within the AR 600–9 standards.

3. This correspondence will be retained in the individual's MPRJ for 36 months from this date.

Signature of Unit Commander

Figure 1. Sample correspondence for Weight Control Program—Continued

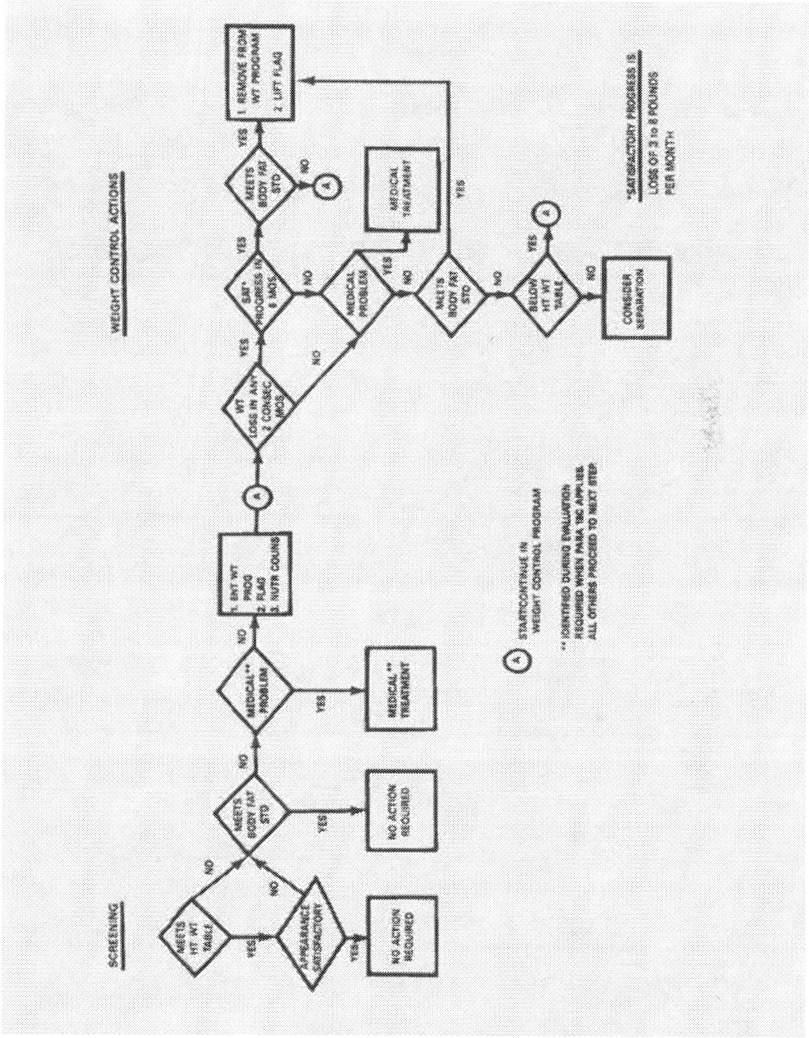

Figure 2. Flow process guide for screening and weight control actions (applies to all personnel; officers, warrant officers, enlisted)

Appendix A
References

Section I
Required Publications

AR 135–175
Separation of Officers. (Cited in paragraph 20*j*.)

AR 135–178
Separation of Enlisted Personnel. (Cited in paragraph 20*j*.)

AR 140–111
Enlistment and Reenlistment. (Cited in paragraph 21*c*.)

AR 600–31
Suspension of Favorable Personnel Actions for Military Personnel in National Security Cases and Other Investigations or Proceedings. (Cited in paragraph 20*e*.)

AR 601–280
Army Reenlistment Program. (Cited in paragraph 21.)

AR 635–100
Officer Personnel. (Cited in paragraph 20*j*.)

AR 635–200
Enlisted Personnel. (Cited in paragraph 20*j*.)

AR 640–10
Individual Military Personnel Records. (Cited in paragraph 20*m*(1).)

NGR 600–5
Enlisted Career Management. (Cited in paragraph 20*j*.)

NGR 600–101
Commissioned Officers—Federal Recognition and Related Personnel Activities. (Cited in paragraph 20*j*.)

NGR 600–200
Enlisted Personnel Management. (Cited in paragraphs 20*j* and 21*d*.)

NGR 635–100
Termination of Appointment and Withdrawal of Federal Recognition. (Cited in paragraph 20*j*.)

Section II
Related Publications
This section contains no entries.

Section III
Prescribed Forms
This section contains no entries.

Section IV
Referenced Forms
This section contains no entries.

Appendix B
Standard Methods for Determining Body Fat Using Body Circumferences, Height and Weight

B–1. Introduction

a. The procedures for the measurements of height, weight, and specific body circumferences for the estimation of body fat are described in this appendix.

b. Although circumferences may be looked upon by untrained personnel as easy measures, they can give erroneous results if proper precautions are not followed. The individual taking the measurements must have a thorough understanding of the appropriate body landmarks and measurement techniques. Unit commanders should require that designated personnel have hands-on training and read the instructions regarding technique and location, and practice before official determinations are made. Two members of the unit should be utilized in the taking of measurements, one to place the tape measure and determine measurements, the other to assure proper placement and tension of the tape, as well as to record the measurement on the worksheet. The individual taking the measurements should be of the same sex as the soldier being measured; the individual who assists the measurer and does the recording may be of either sex. The two should work with the soldier between them so the tape is clearly visible from all sides. Measurements will be made three times, in accordance with standard body measurement procedures. This is necessary for reliability purposes, since the greater number of measurements, the lesser the standard of deviation. Also, if only two measurements were taken, there would be no way to tell which measurement was the most accurate. If there is greater than 1/4-inch difference between the measurements, then continue measuring until you have three measurements within 1/4-inch of each other. An average of the scores that are within 1/4-inch of each other will be used.

c. When measuring circumferences, compression of the soft tissue is a problem that requires constant attention. The tape will be applied so that it makes contact with the skin and conforms to the body surface being measured. It should not compress the underlying soft tissues. Note, however, that in the hip circumference more firm pressure is needed to compress gym shorts. All measurements are made in the horizontal plane, (i.e., parallel to the floor), unless indicated otherwise.

d. The tape measure should be made of a non-stretchable material, preferably fiberglass; cloth or steel tapes are unacceptable. Cloth measuring tapes will stretch with usage and most steel tapes do not conform to body surfaces. The tape measure should be calibrated, i.e., compared with a yardstick or a metal ruler to ensure validity. This is done by aligning the fiberglass tape measure with the quarter inch markings on the ruler. The markings should match those on the ruler; if not, do not use that tape measure. The tape should be 1/4- to 1/2-inch wide (not exceeding 1/2-inch) and a minimum of 5–6 feet in length. A retractable fiberglass tape is the best type for measuring all areas. Tapes currently available through the Army Supply System (Federal Stock Number 8315–00–782–3520) may exceed the 1/2-inch width limits and could slightly impact on circumferential measurements. Efforts are being made to replace the supply system tape with a narrower retractable tape. In the interim, the current Army supply system or any other fiberglass tape not to exceed 5/8-inch may be used if retractable tapes cannot be purchased by unit budget funds available and approved by installation commanders.

B–2. Height and weight measurements

a. The height will be measured with the soldier, in stocking feet (without shoes) and standard PT uniform, i.e., gym shorts and T- shirt, standing on a flat surface with the head held horizontal, looking directly forward with the line of vision horizontal, and the chin parallel to the floor. The body should be straight but not rigid, similar to the position of attention. Unlike the screening table weight this measurement will be recorded to the nearest 1/4-inch in order to gather a more accurate description of the soldier's physical characteristics.

b. The weight will be measured with the soldier in a standard PT uniform, i.e., gym shorts and a T-shirt. Shoes will not be worn. The measurement should be made on scales available in units and recorded to the nearest pound with the following guidelines:

(1) If the weight fraction of the soldier is less than 1/2-pound, round down to the nearest pound.

(2) If the weight fraction of the soldier is 1/2-pound or greater, round up to the next whole pound.

B–3. Description of circumference sites, and their anatomical landmarks and technique

a. All circumference measurements will be taken three times and recorded to the nearest 1/4-inch (or 0.25). Each sequential measurement should be within 1/4-inch of the next or previous measurement. If the measurements are within 1/4-inch of each other, derive a mathematical average to the nearest quarter (1/4) of an inch. If the measurements differ by 1/4-inch or more continue measurements until you obtain three measures within 1/4-inch of each other. Then average the three closest measures.

b. Each set of measurements will be completed sequentially to discourage assumption of repeated measurement readings. For males, complete 1 set of abdomen and neck measurements, NOT three abdomen circumferences followed by three neck circumferences. Continue the process by measuring the abdomen and neck in series until you have three sets of measurements. For females, complete one set of hip, forearm, neck, and wrist measurements, NOT 3 hip followed by three forearm etc. continue the process by measuring hip, forearm, neck, and wrist series until you have 3 sets of measurements.

c. Worksheets for computing body fat are at figure B–1 (males) and figure B–3 (females). Local reproduction is authorized. A blank copy of DA Forms 5500–R and 5501–R is located at the back of this volume. These forms will be reproduced locally on 8½ x 11-inch paper. Supporting factor tables are located at tables B–1 and B–2 (males) and tables B–3 through B–8 (females) and include specific steps for preparing body fat content worksheets.

d. Illustrations of each tape measurement are at figure B–2 (males) and figure B–4 (females). A training videotape (TVT 8–103) is also available at Visual Information Libraries, and/or Training Audiovisual Support Centers (TASC).

B–4. Circumference sites and landmarks for males

a. Abdomen. The soldier being measured will be standing with arms relaxed. The abdominal measurement is taken at a level coinciding with the midpoint of the navel (belly button) with the tape placed so that it is level all the way around the soldier being measured. Record the measurement at the end of a normal expiration. It is important that the soldier does not attempt to hold his abdomen in, thus resulting in a smaller measurement. Also the tape must be kept level across the abdomen and back.

b. Neck. The soldier being measured will be standing, looking straight ahead, chin parallel to the floor. The measurement is taken by placing the tape around the neck at a level just below the larynx (Adam's apple). Do not place the tape measure over the Adam's apple. The tape will be as close to horizontal (the tape line in the front of the neck should be at the same height as the tape line in the back of the neck) as anatomically feasible. In many cases the tape will slant down toward the front of the neck. Therefore, care should be taken so as not to involve the shoulder/ neck muscles (trapezius) in the measurement. This is a possibility when a soldier has a short neck.

B–5. Circumference sites and landmarks for females

a. Neck. This procedure is the same as for males.

b. Forearm. The soldier being measured will be standing with the arm extended away from the body so that the forearm is in plain view of the measurer, with the hand palm up. The soldier should be allowed to choose which arm he/she prefers to be measured. Place the tape around the largest forearm circumference. This will be just below the elbow. To ensure that this is truly the largest circumference, since it is being visually identified, slide the tape along the forearm to find the largest circumference.

c. Wrist. The soldier being measured will stand with the arm extended away from the body so that the wrist is in plain view of the measurer. The tape will be placed around the wrist at a point above the hand just below the lower end of the bones of the forearm.

d. Hip. The soldier taking the measurement will view the person being measured from the side. Place the tape around the hips so that it passes over the greatest protrusion of the gluteal muscles (buttocks) keeping the tape in a horizontal plane (i.e., parallel to the floor). Check front to back and side to side to be sure the tape is level to the floor on all sides before the measurements are recorded. Since the soldier will be wearing gym shorts, the tape can be drawn snugly to minimize the influence of the shorts on the size of the measurement.

B–6. Preparation of the body fat content worksheets

NOTE: IT IS EXTREMELY IMPORTANT THAT YOU READ ALL OF THESE INSTRUCTIONS BEFORE ATTEMPTING TO COMPLETE THE BODY FAT CONTENT WORKSHEETS. MAKE SURE THAT YOU HAVE A COPY OF THE WORKSHEET IN FRONT OF YOU WHEN YOU ARE READING THESE INSTRUCTIONS.

a. The following paragraphs will provide information needed to perpare the Body Fat Content Worksheets for males and females, DA Form 5500–R and 5501–R, Dec 85. The worksheets are written in a stepwise fashion. The measurements and computation processes are different for males and females.

b. You will be responsible for completing a worksheet for soldiers who exceed the screening table weight (Table 1) located in this regulation, or when a unit commander or supervisor determines that the individual's appearance suggests that body fat is excessive (para 20*a* AR 600–9). The purpose of this form is to help you determine the soldier's percent body fat using the circumference technique described in this regulation.

c. Before you start, you should have a thorough understanding of the measurements to be made as outlined in this appendix. You will also need a scale for measuring body weight, a height measuring device, and a measuring tape (see specifications in para B–1*d*) for the circumference measurements.

B–7. Steps for preparing the Male Body Fat Content Worksheet, DA Form 5500–R, Dec 85

Name Print the soldier's last name, first name, and middle initial in the NAME block. Also include his Rank, and Social Security Number.

Age Print his age in years in the AGE block.

Height Measure the soldier's height as described in this appendix, to the nearest quarter of an inch, and record the measurement in the HEIGHT block.

Weight Measure the soldier's weight as described in this appendix, to the nearest pound, and record in the WEIGHT block.

Note: Follow the rules for rounding of height and weight measurements as described earlier in this appendix.

Step 1. Abdominal Measurement

Measure the soldier's abdominal circumference to the nearest quarter of an inch, and record in the block labeled "FIRST".

Step 2. Neck Measurement

Measure the soldier's neck circumference to the nearest quarter of an inch, and record in the block labeled "FIRST".

Note: REPEAT STEPS 1 and 2, in series until you have completed three sets of abdomen and neck circumferences.

Step 3. Average Abdominal Measurement

Find the mathematical average of the FIRST, SECOND, and THIRD abdominal circumferences by adding them together and dividing by three. Place this number to the nearest quarter of an inch, in the block marked AVERAGE, for STEPS 1 and 3.

Step 4. Average Neck Measurement

Find the mathematical average of your FIRST, SECOND, and THIRD neck circumferences by adding them together and dividing by three. Place this number to the nearest quarter of an inch in the block marked AVERAGE, for STEPS 2 and 4.

Step 5. Abdomen-Neck Difference

Subtract the number found in the AVERAGE block of STEP 4 from the number found in the AVERAGE block in STEP 3. Enter the result in STEP 5. This is the difference between the abdomen and neck circumferences.

Step 6. Abdomen-Neck Factor

Go to Table B–1, the Abdomen-Neck Factor Table, and locate the abdomen-neck difference in the left-most column. If the difference is a whole number, i.e., 15 inches, the Abdomen-Neck Factor is 89.93. If the difference is 15.25 inches, the factor would be 90.48, if the difference is 15.50, the factor is 91.02, and if the difference is 15.75, the factor is 91.55. Enter the appropriate factor in STEP 6.

Step 7. Height Factor

Go to Table B–2, the Height Factor Table, and locate the soldier's height in the left-most column. If the height is a whole number, i.e., 64 inches, the factor is 77.15. If the height is not a whole number, i.e., 64.25 inches, the factor is 77.27, if the height is 64.50 inches the factor is 77.39, and if the height is 64.75 inches, the factor is 77.50. Enter the appropriate factor in STEP 7.

Note: Therefore, the general rule for the factor tables is if the measure or difference is a whole number, your factor will be located under the 0.00 column, directly across from the inches column. Columns .25, .50, .75 correspond to measurements that are not whole numbers, but rather are fractions of an inch.

Step 8. Percent Body Fat

Subtract the number found in the AVERAGE block of STEP 7 from the number found in the AVERAGE block of STEP 6 and enter the difference in STEP 8. This is the soldier's PERCENT BODY FAT.

Table B–1
Male Abdomen and Neck Factor

Difference in inches	0.00	.25 (¼)	.50 (½)	.75 (¾)
5	53.44	55.06	56.61	58.09
6	59.50	60.85	62.16	63.41
7	64.62	65.78	66.91	68.00
8	69.05	70.07	71.07	72.03
9	72.96	73.87	74.76	75.62
10	76.46	77.28	78.08	78.86
11	79.63	80.37	81.10	81.82
12	82.52	83.20	83.87	84.53
13	85.17	85.81	86.43	87.04
14	87.64	88.22	88.80	89.37
15	89.93	90.48	91.02	91.55
16	92.07	92.58	93.09	93.59
17	94.08	94.57	95.05	95.52
18	95.98	96.44	96.89	97.34
19	97.78	98.21	98.64	99.06
20	99.48	99.89	100.30	100.70
21	101.10	101.49	101.88	102.26
22	102.64	103.02	103.39	103.76
23	104.12	104.48	104.83	105.19
24	105.53	105.88	106.22	106.56
25	106.89	107.22	107.55	107.87
26	108.19	108.51	108.82	109.14
27	109.44	109.75	110.05	110.35
28	110.65	110.95	111.24	111.53
29	111.82	112.10	112.39	112.67
30	112.94	113.22	113.49	113.76
31	114.03	114.30	114.56	114.83
32	115.09	115.35	115.60	115.86
33	116.11	116.36	116.61	116.85
34	117.10	117.34	117.58	117.82
35	118.06	118.30	118.53	118.77
36	119.00	119.23	119.46	119.68
37	119.91	120.13	120.35	120.57
38	120.79	121.01	121.23	121.44
39	121.66	121.87	122.08	122.29
40	122.50	122.70	122.91	123.11

Table B-2
Male Height Factor

Inches	0.00	.25 (¼)	.50 (½)	.75 (¾)
60	75.23	75.35	75.48	75.60
61	75.72	75.84	75.96	76.09
62	76.21	76.33	76.45	76.56
63	76.68	76.80	76.92	77.04
64	77.15	77.27	77.39	77.50
65	77.62	77.73	77.84	77.96
66	78.07	78.18	78.30	78.41
67	78.52	78.63	78.74	78.85
68	78.96	79.07	79.18	79.29
69	79.40	79.50	79.61	79.72
70	79.83	79.93	80.04	80.14
71	80.25	80.35	80.46	80.56
72	80.67	80.77	80.87	80.98
73	81.08	81.18	81.28	81.38
74	81.48	81.58	81.68	81.78
75	81.88	81.98	82.08	82.18
76	82.28	82.38	82.47	82.57
77	82.67	82.77	82.86	82.96
78	83.05	83.15	83.24	83.34
79	83.43	83.53	83.62	83.72
80	83.81	83.90	83.99	84.09
81	84.18	84.27	84.36	84.45
82	84.54	84.64	84.73	84.82
83	84.91	85.00	85.08	85.17
84	85.26	85.35	85.44	85.53

B–8. Steps for preparing the Female Body Fat Content Worksheet, DA Form 5501–R, Dec 85

Name Print the soldier's last name, and middle initial in the NAME block. Also include her Rank, and Social Security Number.

Age Print her age in years in the AGE block.

Height Measure the soldier's height as described in this appendix, to the nearest quarter of an inch, and record the measurement in the HEIGHT block.

Weight Measure the soldier's weight as described in this appendix, to the nearest pound, and record in the WEIGHT block.

Note: Follow the rounding rules for rounding height and weight measurements as described earlier in this appendix.

Step 1. Weight Factor

Go to Table B–3, the Weight Factor Table, and locate the soldier's weight in the left-most column, which is in 10 pound increments. If the weight is exactly 120 pounds, the factor is found under the "0" column and is 147.24. If the weight is 121 pounds, the factor is found under the "1" column and is 147.62. If the weight is 126 the factor is found under the "6" column and is 149.47. Enter the appropriate weight factor in the CALCULATIONS section, STEP 11 A.

Step 2. Height Factor

Go to Table B–4, the Height Factor Table, and locate the soldier's height in the left-most column. If the height is a whole number, i.e., 64 inches, the factor is found under the 0.00 column and is 83.75. If the height is not a whole number, i.e., 64.25 inches, the factor is 84.07. If the height is 64.50 inches, the factor is 84.40, and if the height is 64.75 inches, the factor is 84.73. Enter the appropriate height factor in the CALCULATIONS section, STEP 11 D.

Step 3. Hip Measurement

Measure the soldier's hip circumference to the nearest quarter of an inch, and record in the block labeled "FIRST."

Step 4. Forearm Measurement

Measure the soldier's forearm to the nearest quarter of an inch, and record in the block labeled "FIRST."

Step 5. Neck Measurement

Measure the soldier's neck circumference to the nearest quarter of an inch, and record in the block labeled "FIRST."

Step 6. Wrist Measurement

Measure the soldier's wrist to the nearest quarter of an inch, and record in the block labeled "FIRST."

Note: REPEAT STEPS 3, 4, 5, and 6, IN SERIES, until you have completed 3 sets of Hip, Forearm, Neck and Wrist circumferences. When you have completed this series, find the mathematical average for *each* of the 4 circumference measures and place each average in its respective AVERAGE block.

Step 7. Hip Factor

Go to Table B–5, the Hip Factor Table, and locate the soldier's AVERAGE hip circumference in the left-most column. If the circumference is a whole number, i.e., 36 inches, the Hip Factor is found in the 0.00 column and is 15.83. If the circumference is not a whole number but is 36.25 inches, the factor is 15.94. If the circumference is 36.50 the factor is 16.05. Enter the appropriate factor in the CALCULATIONS section, 11 B.

Step 8. Forearm Factor

Go to Table B–6, the Forearm Factor Table, and locate the soldier's AVERAGE forearm circumference in the left-most column. If the circumference is a whole number, i.e., 10 inches, the factor is found under 0.00 column and is 39.97. If the circumference is not a whole number but is 10.25 inches, the factor is 40.97. If the circumference is 10.75 inches, the factor is 40.97. Enter the appropriate factor in the CALCULATIONS, 11 E.

Step 9. Neck Factor

Go to Table B–7, the Neck Factor Table, and locate the soldier's AVERAGE neck circumference in the left-most column. If the circumference is a whole number, i.e., 12 inches, the factor is found under the 0.00 column and is 16.25. If the circumference is not a whole number but is 12.25 inches, the factor is 16.59. If the circumference is 12.50 inches, the factor is 16.93. If the circumference is 12.75 inches, the factor is 17.26. Enter the appropriate factor is the CALCULATIONS section, 11 F.

Step 10. Wrist Factor

Go to Table B–8, the Wrist Factor Table, and locate the soldier's AVERAGE wrist circumference in the left-most column. If the circumference is a whole number, i.e., 7 inches, the factor is found under the 0.00 column and is 3.56. If the circumference is not a whole number but is 7.25 inches, the factor is 3.69. If the circumference is 7.50 inches, the factor is 3.82. If the circumference is 7.75 inches, the factor is 3.94. Enter the appropriate factor in the CALCULATIONS section, 11 G.

Calculations

Line C. Addition of Weight and Hip Factors

Add 11 A, Weight Factor, to 11 B, Hip Factor. Enter the result on line 11 C (Total).

Line H. Addition of Height, Forearm, Neck, and Wrist Factors

Add 11 D, Height Factor, 11 E, Forearm Factor, 11 F, Neck Factor, and 11 G, Wrist Factor together. Enter the result on line 11 H. (Total).

Line I. Percent Body Fat

Subtract Line 11–H from Line–C and enter on Line I. This is the soldier's PERCENT BODY FAT.

Table B–3
Female Weight Factor

Pounds	0	1	2	3	4	5	6	7	8	9
90	134.08	134.58	135.08	135.58	136.07	136.55	137.03	137.50	137.97	138.44
100	138.90	139.35	139.80	140.25	140.69	141.13	141.56	141.99	142.42	142.84
110	143.26	143.67	144.08	144.49	144.89	145.29	145.69	146.08	146.47	146.85
120	147.24	147.62	147.99	148.37	148.74	149.10	149.47	149.83	150.19	150.54
130	150.90	151.25	151.60	151.94	152.28	152.62	152.96	153.30	153.63	153.96
140	154.29	154.61	154.94	155.26	155.58	155.89	156.21	156.52	156.83	157.14
150	157.44	157.75	158.05	158.35	158.65	158.94	159.24	159.53	159.82	160.11
160	160.40	160.68	160.96	161.25	161.53	161.80	162.08	162.35	162.63	162.90
170	163.17	163.44	163.70	163.97	164.23	164.49	164.76	165.01	165.27	165.53
180	165.78	166.04	166.29	166.54	166.79	167.04	167.28	167.53	167.77	168.02
190	168.26	168.50	168.74	168.97	169.21	169.44	169.68	169.91	170.14	170.37
200	170.60	170.83	171.06	171.28	171.51	171.73	171.96	172.18	172.40	172.62
210	172.83	173.05	173.27	173.48	173.70	173.91	174.12	174.33	174.54	174.75
220	174.99	175.20	175.41	175.61	175.82	176.02	176.22	176.42	176.62	176.82
230	177.03	177.22	177.42	177.62	177.81	178.00	178.20	178.40	178.59	178.82
240	178.97	179.16	179.35	179.54	179.73	179.92	180.10	180.29	180.47	180.66
250	180.84	181.02	181.20	181.39	181.57	181.75	181.92	182.10	182.28	182.46
260	182.63	182.81	182.98	183.16	183.33	183.51	183.68	183.85	184.02	184.19
270	184.36	184.53	184.70	184.87	185.03	185.20	185.37	185.53	185.70	185.86

Table B–4
Female Height Factor

Inches	0.00	.25 (¼)	.50 (½)	.75 (¾)
55	71.97	72.30	72.62	72.95
56	73.28	73.61	73.93	74.26
57	74.59	74.91	75.24	75.57
58	75.90	76.22	76.55	76.88
59	77.20	77.53	77.86	78.19
60	78.51	78.84	79.17	79.49
61	79.82	80.15	80.48	80.80
62	81.13	81.46	81.78	82.11
63	82.44	82.77	83.09	83.42
64	83.75	84.07	84.40	84.73
65	85.06	85.38	85.71	86.04

Table B–4
Female Height Factor—Continued

Inches	0.00	.25 (¼)	.50 (½)	.75 (¾)
66	86.36	86.69	87.02	87.35
67	87.67	88.00	88.33	88.65
68	88.98	89.31	89.64	89.96
69	90.29	90.62	90.94	91.27
70	91.60	91.93	92.25	92.58
71	92.91	93.23	93.56	93.89
72	94.22	94.54	94.87	95.20
73	95.52	95.85	96.18	96.51
74	96.83	97.16	97.49	97.81
75	98.14	98.47	98.80	99.12
76	99.45	99.78	100.10	100.43
77	100.76	101.09	101.41	101.74
78	102.07	102.39	102.72	103.05
79	103.38	103.70	104.03	104.36
80	104.68	105.01	105.34	105.67
81	105.99	106.32	106.65	106.97
82	107.30	107.63	107.96	108.28
83	108.61	108.94	109.26	109.59
84	109.92	110.24	110.57	110.90

Table B–5
Female Hip Factor

Inches	0.00	.25 (¼)	.50 (½)	.75 (¾)
30	13.19	13.30	13.41	13.52
31	13.63	13.74	13.85	13.96
32	14.07	14.18	14.29	14.40
33	14.51	14.62	14.73	14.84
34	14.95	15.06	15.17	15.28
35	15.39	15.50	15.61	15.72
36	15.83	15.94	16.05	16.15
37	16.26	16.37	16.48	16.59
38	16.70	16.81	16.92	17.03
39	17.14	17.25	17.36	17.47
40	17.58	17.69	17.80	17.91
41	18.02	18.13	18.24	18.35
42	18.46	18.57	18.68	18.79
43	18.90	19.01	19.12	19.23
44	19.34	19.45	19.56	19.67
45	19.78	19.89	20.00	20.11
46	20.22	20.33	20.44	20.55
47	20.66	20.77	20.88	20.99
48	21.10	21.21	21.32	21.43
49	21.54	21.65	21.76	21.87
50	21.98	22.09	22.20	22.31
51	22.42	22.53	22.64	22.75
52	22.86	22.97	23.08	23.19
53	23.30	23.41	23.52	23.63

Table B–6
Female Forearm Factor

Inches	0.00	.25 (¼)	.50 (½)	.75 (¾)
5	19.98	20.98	21.98	22.98
6	23.98	24.98	25.98	26.98
7	27.98	28.98	29.98	30.97
8	31.97	32.97	33.97	34.97
9	35.97	36.97	37.97	38.97
10	39.97	40.97	41.97	42.97
11	43.96	44.96	45.96	46.96
12	47.96	48.96	49.96	50.96
13	51.96	52.96	53.96	54.96
14	55.95	56.95	57.95	58.95
15	59.95	60.95	61.95	62.95

Table B–7
Female Neck Factor

Inches	0.00	.25 (¼)	.50 (½)	.75 (¾)
5	6.77	7.11	7.45	7.79
6	8.12	8.46	8.80	9.14
7	9.48	9.82	10.16	10.49
8	10.83	11.17	11.51	11.85
9	12.19	12.53	12.86	13.20
10	13.54	13.88	14.22	14.56
11	14.90	15.23	15.57	15.91
12	16.25	16.59	16.93	17.26
13	17.60	17.94	18.28	18.62
14	18.96	19.30	19.63	19.97
15	20.31	20.65	20.99	21.33

Table B–8
Female Wrist Factor

Inches	0.00	.25 (¼)	.50 (½)	.75 (¾)
5	2.54	2.67	2.80	2.93
6	3.05	3.18	3.31	3.43
7	3.56	3.69	3.82	3.94
8	4.07	4.20	4.33	4.45
9	4.58	4.71	4.83	4.96
10	5.09	5.22	5.34	5.47
11	5.60	5.72	5.85	5.98
12	6.11	6.23	6.36	6.49
13	6.62	6.74	6.87	7.00
14	7.12	7.25	7.38	7.51
15	7.63	7.76	7.89	8.01

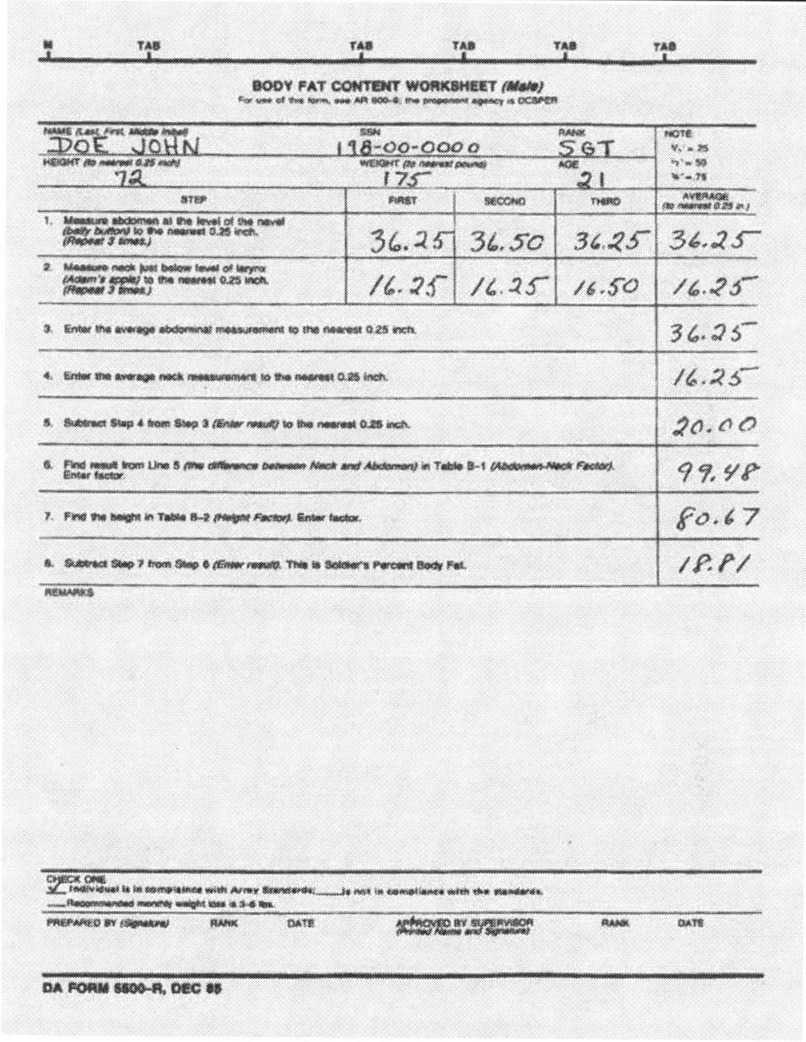

M TAB TAB TAB TAB TAB

BODY FAT CONTENT WORKSHEET *(Male)*

For use of this form, see AR 600-9; the proponent agency is DCSPER

NAME *(Last, First, Middle Initial)*	SSN	RANK	NOTE
DOE JOHN	198-00-0000	SGT	¼" = .25
HEIGHT *(to nearest 0.25 inch)*	WEIGHT *(to nearest pound)*	AGE	½" = .50
72	175	21	¾" = .75

STEP	FIRST	SECOND	THIRD	AVERAGE *(to nearest 0.25 in.)*
1. Measure abdomen at the level of the navel *(belly button)* to the nearest 0.25 inch. *(Repeat 3 times.)*	36.25	36.50	36.25	36.25
2. Measure neck just below level of larynx *(Adam's apple)* to the nearest 0.25 inch. *(Repeat 3 times.)*	16.25	16.25	16.50	16.25
3. Enter the average abdominal measurement to the nearest 0.25 inch.				36.25
4. Enter the average neck measurement to the nearest 0.25 inch.				16.25
5. Subtract Step 4 from Step 3 *(Enter result)* to the nearest 0.25 inch.				20.00
6. Find result from Line 5 *(the difference between Neck and Abdomen)* in Table B-1 *(Abdomen-Neck Factor)*. Enter factor.				99.48
7. Find the height in Table B-2 *(Height Factor)*. Enter factor.				80.67
8. Subtract Step 7 from Step 6 *(Enter result)*. This is Soldier's Percent Body Fat.				18.81

REMARKS

CHECK ONE

✓ Individual is in compliance with Army Standards; ____ is not in compliance with the standards.

____ Recommended monthly weight loss is 3-6 lbs.

PREPARED BY *(Signature)*	RANK	DATE	APPROVED BY SUPERVISOR *(Printed Name and Signature)*	RANK	DATE

DA FORM 5500-R, DEC 85

Figure B–1. Sample of a completed DA Form 5500

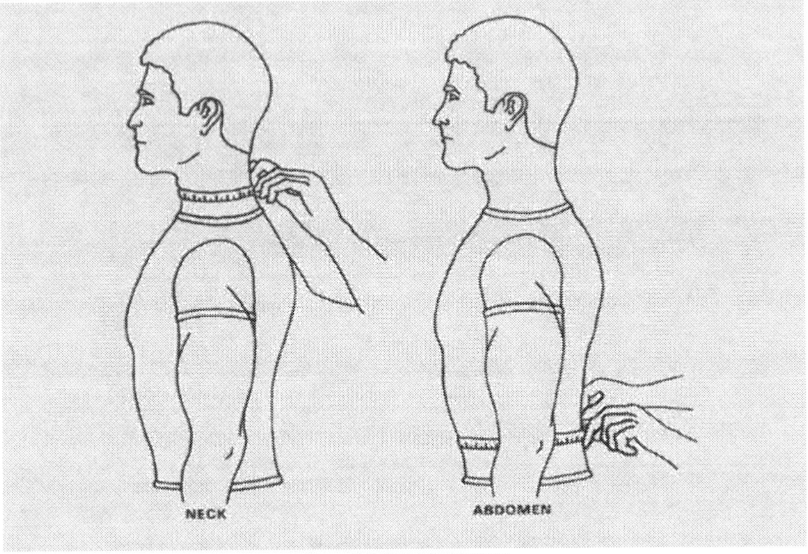

Figure B–2. Male Measurement

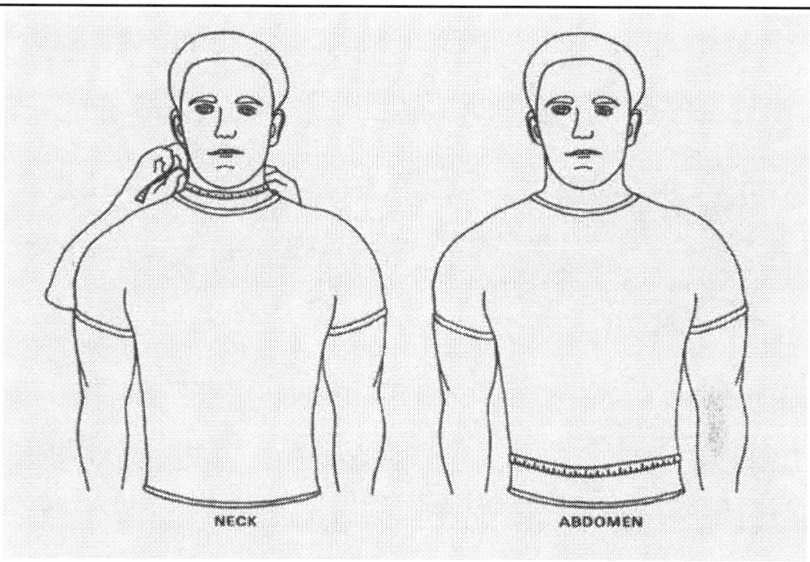

Figure B–2. Male Measurement—Continued

M — TAB — TAB — TAB — TAB

BODY FAT CONTENT WORKSHEET *(Female)*

For use of this form, see AR 600-9; the proponent agency is DCSPER

NAME *(Last, First, Middle initial)*	SSN	RANK	NOTE:
DOE, JANE	432-00-0000	SGT	¼" = .25
HEIGHT *(to nearest 0.25 inch)*	WEIGHT *(to nearest pound)*	AGE	½" = .50
63.00	132	21	¾" = .75

STEP	FIRST	SECOND	THIRD	AVERAGE *(to nearest 0.25 in.)*
1. Find the soldier's weight in Table B-3 *(Weight Factor)*. Enter factor in 11A below				
2. Find soldier's height in Table B-4 *(Height Factor)*. Enter factor in 11D below.				
3. Measure hips at point where the gluteus muscles *(buttocks)* protrude backward the most. Round off to nearest 0.25 inch. Repeat three times, then average.	29.75	30.00	30.25	30.00
4. Measure forearm at its largest point *(with arm horizontal, palm up)* to nearest 0.25 inch. Repeat three times, then average.	8.25	8.50	8.75	8.50
5. Measure neck just below level of larynx *(Adam's apple)* to nearest 0.25 inch. Repeat three times and average.	13.25	13.50	13.75	13.50
6. Measure wrist between the bones of the hand and forearm to nearest 0.25 inch. Repeat three times, then average.	5.75	6.00	6.25	6.00
7. Find average hip measurement in Table B-5 *(Hip Factor)*. Enter factor in 11B below.				
8. Find average forearm measurement in Table B-6 *(Forearm Factor)*. Enter factor in 11E below.				
9. Find average neck measurement in Table B-7 *(Neck Factor)*. Enter factor in 11F below.				
10. Find average wrist measurement in Table B-8 *(Wrist Factor)*. Enter factor in 11G below.				

11. CALCULATIONS			REMARKS
A. Weight factor	151.60		
B. Hip factor	13.19		
C. TOTAL *(11A+11B)*		164.79	
D. Height factor	82.44		
E. Forearm factor	33.97		
F. Neck factor	18.28		
G. Wrist factor	3.05		
H. TOTAL *(11D+E+F+G)*		137.74	
I. SOLDIER'S PERCENT BODY FAT *(Line 11C—11H)*		27.05	

CHECK ONE
✓ Individual is in complaince with Army Standards; _____ is not in compliance with the standards.
____ Recommended monthly weight loss is 3-8 lbs.

PREPARED BY *(Signature)*	RANK	DATE	APPROVED BY SUPERVISOR *(Printed Name and Signature)*	RANK	DATE

DA FORM 5501-R, DEC 85

Figure B–3. Sample of a completed DA Form 5501

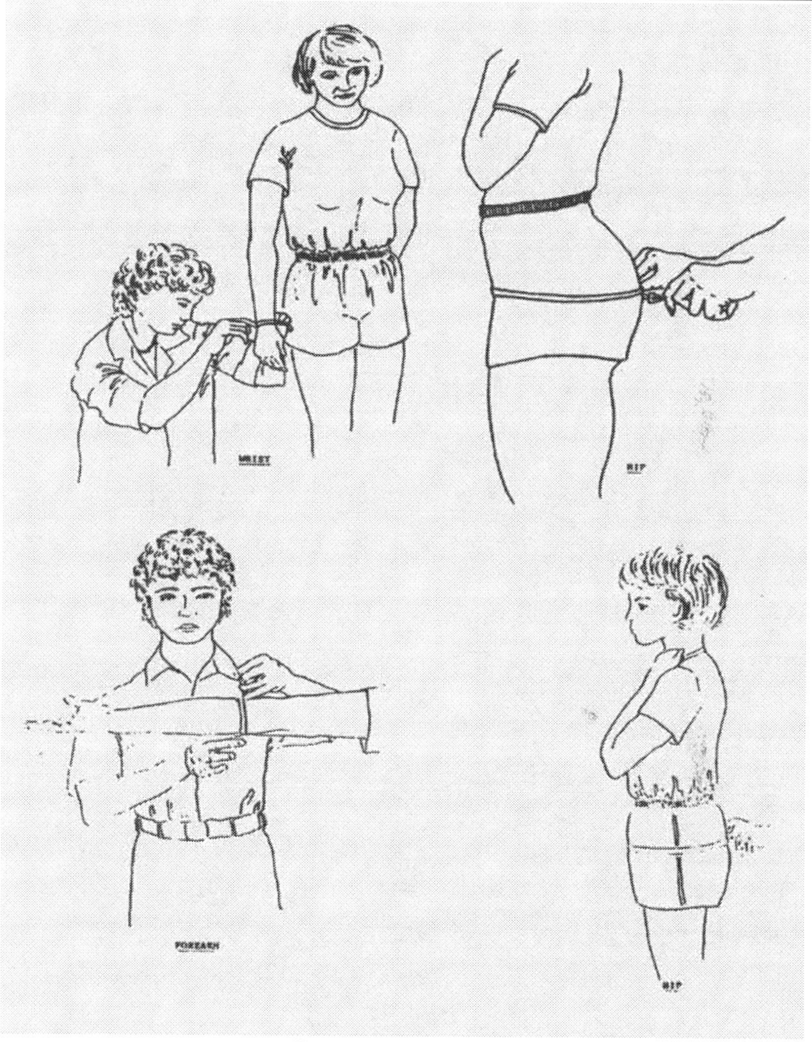

Figure B–4. Female Measurement

Appendix C
Nutrition Guide to the Weight Control Program

C–1. General
This appendix explains the basic principles of weight loss while maintaining normal nutrition. It does not replace the requirement for Active and Reserve Component Units with soldiers exceeding the body fat standards to be provided weight reduction counseling by qualified health care personnel. This guide should be used as a supplement to weight reduction counseling and as a guide for commanders in developing an effective weight control program.

C–2. Weight control—it's time to make the fitness connection
In many cases fitness begins with weight reduction. Aerobic fitness is related to an individual's body fat. The higher the fat, the less likely the individual is to be aerobically fit and the harder it is to maintain higher levels of physical stamina and endurance. A fit Army is a lean Army. Military readiness demands weight control—the spin-off is less likelihood of developing heart disease, high blood pressure, and diabetes. Proper nutrition and regular exercise are necessary to help you lose weight and improve your state of fitness.

a. Invest in yourself.

(1) Make a decision to lose weight and shape up.

(2) Get motivated.

(3) Develop a strategy (diet, exercise routine, lifestyle changes, etc.).

(4) Carry out the strategy.

b. Enjoy the payoff.

(1) A healthy appearance.

(2) An improved self image.

(3) A sense of accomplishment.

(4) A feeling of pride.

C–3. Making nutrition work
Improving your nutrition will increase your mileage in many ways. You can even lose weight while improving your performance. Your nutrition program should include the right number of calories to cause a steady loss of body fat with no loss of energy. Stay away from food fads—they're usually boring, unhealthy, too strict and lead only to temporary weight loss. Compare the benefits of a sound nutrition with the consequences of crash diets as shown below:

Table C–0–1
Nutrition

Sound Nutrition: Provides all required nutrients **Crash Diets:** Most often lack some nutrients
Sound Nutrition: Gradual loss of body fat (1 to 2 pounds per week) **Crash Diets:** Rapid loss of body water and muscle mass—not body fat
Sound Nutrition: Reinforces a good mental outlook **Crash Diets:** Symptoms of grumpiness, headaches, anxiety and fatigue
Sound Nutrition: Improves health **Crash Diets:** Can cause depression, dehydration, potential serious illness and a slow-down in your body's metabolism (the rate your body burns calories)
Sound Nutrition: Gives a sense of accomplishment **Crash Diets:** End in eventual weight loss and failure
Sound Nutrition: Develops permanent good eating habits **Crash Diets:** Encourage unhealthy eating habits of temporary duration

C–4. Basic strategy checklist—a plan for making the right connection

a. Good attitude. Having a good mental attitude is necessary to succeed in any program. To lose weight, a good mental attitude helps self-discipline—an important ingredient.

b. Sensible nutrition. A diet of adequate essential nutrients is necessary to prevent mental and physical fatigue. Crash diets don't work in a permanent weight control program.

c. Regular exercise. Exercise promotes physical fitness. It improves flexibility, strength, endurance and weight loss by speeding up the body's metabolism. It has also been proven to help supress the appetite.

d. Diet and exercise master plan. Create your own daily food intake and exercise plan. Keep a record. Make adjustments. You are in control.

e. Rest. Adequate rest improves attitude, posture, and appearance. Lack of rest and sleep can weaken resistance and will power.

f. A set goal. See a long range objective, then get going by setting easier-to-reach short-term goals. After the first few goals are met, the objective will be in plain view.

C–5. Balance check

If you need to lose weight, it's time to check your energy balance to see what's tilting the scale. When your energy input (calories) is greater than your energy output (activity), you store the extra calories as body fat. To stay in balance:

a. Increase exercise (frequency and intensity), or

b. Decrease calorie intake, or

c. Combine exercise with caloric reduction for best results.

C–6. Obesity risks

Excess body fat is harmful to your health. It increases your risk for developing high blood pressure, diabetes, heart disease, respiratory infections, gall bladder disease, low-back pain, and some forms of cancer. It has further drawbacks in physical appearance and interferes with physical performance.

C–7. Tailoring a nutrition program

The best nutrition program is one that allows you to lose body fat while you eat regular wholesome foods in controlled portion sizes. Does your diet include essential nutrients? Here's an easy way to check: (See below).

Table C–0–2
Diet structures

Food group	Recommended Servings Per day	Your diet
Milk (8 ozs)	2	—
Meat (3 ozs)	2	—
Bread or Cereal (1 slice or ½ cup)	4	—
Vegetables ½ cup)	2	—
Fruits (1 medium or 1/2 cup)	2	—
Does your diet provide the right amount of calories?		

Table C–0–3
Weight maintenance formula

__________	X	__________	=	__________
Your ideal body weight		Activity factor		Your calorie level for maintenance

a. Weight maintenance.

b. Activity factors.

(1) *Sedentary.* Twelve to 14 calories per pound are required if you are not involved in exercise on the job or off duty.

(2) *Active.* Fifteen calories per pound are required when your job involves physical work and/or you are engaged in a regular exercise program.

(3) *Highly active.* Sixteen to 18 calories are required per pound when very physically demanding work and/or high level of physical training is done routinely. (Most people do not fit in this category).

c. Weight loss.

(1) To lose 1 pound of fat per week, subtract 500 calories per day from your calorie maintenance level.

(2) To lose 2 pounds of fat per week, subtract 1,000 calories per day. For example:

Table C–0–4
Weight loss formula

________________	less	________________	=	________________
Calorie level for maintenance. (A above)		Calories to meet your weekly goal. (B above)		Your daily calorie level for weight loss.

C–8. What about calories

Calories don't deserve a bad name unless your intake is greater than your requirement. Calories are provided by certain nutrients in the foods and beverages you consume.

a. The following nutrients provide calories:

Table C–0–5
Nutrient-calories guide

Nutrient: Carbohydrate (Provides an efficient fuel source for the body.)
Calories: 4 per gram

Nutrient: Protein (Provides material to repair and build tissues.)
Calories: 4 per gram

Nutrient: Fat (Provides essential fatty acids and concentrated energy source for the body.)
Calories: 9 per gram

Notes:
Alcohol, while not a nutrient does provide calories (7 per gram)

b. Listed in the table below are the nutrients that do not provide calories.

Table C–0–6
Vitamins–Minerals–Water

Nutrient: Vitamins (Needed to utilize the food you eat.)
Calories: 0

Nutrient: Minerals (Needed for bones, teeth, and chemical functions.)
Calories: 0

Nutrient: Water (Necessary for life.)
Calories: 0

C–9. Good nutrition—a personal choice

The amount of calories you consume depends on the type of food you choose, its preparation and the amount you eat (PORTION SIZE). Some foods are very concentrated in calories, (i.e., the portion size is small for the amount of calories it contains). Examples are fats, candy, fried foods, most deserts, and alcohol. These foods are also of low nutrient density—they provide few nutrients for the amount of calories provided. (See below for comparison.)

Table C–0–7
High calorie-low nutrient foods

	Calories
1 piece of pecan pie	750
1 large apple	80
One-half *fried* chicken breast	400
One-half *baked* chicken breast	180
1 potato, French *fried*	275
1 *baked* potato	100
ONE-HALF CUP *syrup pack* peaches	100
One-half cup *unsweetened* peaches	50

C–10. Easy ways to save calories

One sure way to save calories without reducing PORTION SIZE is to choose foods that are:

a. Lower is sugar. (Each teaspoon supplies an additional 20 calories).

b. Lower in fat. (Each teaspoon supplies an additional 45 calories).

c. See table C–1 and table C–2 for easy ways to save calories. The Right Image of Me in the Military (TRIMM) sample meals are lower in fat and sugar—thus more "nutrient dense" and lower in calories. See table C–3 for the 1200 calorie Daily Menu and table C–4 for the 1500-calorie Daily Menu.

C–11. Portion control

a. You'll need to learn to correctly estimate portion sizes in order to ensure adequate nutrition and to control intake. A guide to estimate portion sizes is shown below.
2 ounces=1 slice meat 3 x 4 inches and 1/4-inch thick (cooked)
3 ounces=1 meat pattie or portion (1/5 of a pound)
3-inch diameter, 1/2-inch thick (cooked)
3 ounces=1/2 small chicken breast
1 ounce=1 small chicken drumstick
2 ounces=1 chicken thigh
1 ounce=1/4 cup chopped meat (tuna, spagetti meatsauce, chili, ground meat)

b. Some equal measurements are as follows:

Table C–0–8
Equal measurements

1 cup	=	8 ounces
1 tablespoon	=	3 teaspoons
1/4 cup	=	4 tablespoons
1 ounce	=	40 grams
1 liter	=	34 ounces
.7 liter	=	24 ounces
.5 liter	=	17 ounces
.2 liter	=	7 ounces

c. There is no single food that is so high in calories that a small amount cannot be eaten occasionally. Many people, however, have a particular food obsession that must be recognized. For them trying to eat "just a cookie, piece of candy, or sparerib" is too tempting. The urge to eat "the whole thing" becomes too great. You have to make and follow your own rules according to your ability to control what you eat. Avoidance is one means of control. But if you plan your diet, and diet according to your plan, you can include a favorite high caloric food item as a special occasional treat.

C–12. "Good cookin' for good lookin'"— A memo to the cook for cutting calories during food preparation

a. The milk and cheese group.

(1) Use skim or lowfat milk in recipes when making puddings, sauces, soups, and baked products.

(2) Substitute plain, unsweetened lowfat yogurt or blenderized lowfat cottage cheese in recipes that call for sour cream or mayonaise.

b. The meat, poultry, fish, and dry beans group.

(1) TRIM fat from meat. Cook meats on rack so that fat can drain off.

(2) Roast, bake, broil, or simmer meat, poultry or fish without adding fat. Braise in covered pan on stove top or pan broil in a nonstick pan; and add spices to enhance flavors.

(3) Remove skin from chicken or turkey.

(4) Chill meat broth until fat turns light and solid on top. With a spoon or knife, skim or peel fat off and discard.

c. The vegetable and fruit group.

(1) Steam, boil, broil, or bake vegetables. Some fruits may be broiled or heated with spices added for flavor.

(2) Go easy on sauces, butter, and margarine. Season with herbs and spices. Crisp-cooked vegetables usually don't require as much seasoning as over-cooked vegetables.

(3) Try lemon juice or vinegar on salads. Cut way back on regular salad dressings. (One-fourth cup creamy dressing is approximately 340 calories!)

(4) Read nutrition information labels on food packages.

d. The bread and cereal group.

(1) Use less fat and sugar than called for in recipes. Substitute lower calorie ingredients.

(2) Avoid recipes for baked products that require large amounts of fat and sugar.

(3) Check ingredient labels for fat and sugar content. Check nutrition information label for total calories in each portion.

(4) Use diet margarine or plain yogurt on baked potatoes instead of margarine, butter, or sour cream.

(5) Have boiled, steamed, or baked rather than fried potatoes.

C–13. Dining tips

a. Avoid gravies, sauces, and deep-fried food. If the meat has been fried (southern style chicken, schnitzel), remove the coating and eat only the meat.

b. Remove all the visible fat from the meat.

c. Request diet salad dressing, vinegar, or lemon juice for your salad; most restaurants have them.

d. Starchy foods are not fattening when consumed in moderate quantities. However, avoid those prepared in cream sauces or deep-fried. For example, baked potatoes with a small amount of sour cream or margarine is a good choice. Also, noodles, rice, macaroni, or spaghetti are good potato substitutes.

e. Avoid rich desserts, ice cream, jello, pastry, candy, cookies, pies, cakes, sugar, honey, jam, jelly, regular soda, and other sweets. These are sources of concentrated calories that quickly cause your total intake to skyrocket in just a few bites. USE SPARINGLY, if you must.

f. If you MUST have a snack, have fresh fruits, a few crackers or pretzels, or delicious low-calorie raw vegetables.

g. Low-calorie beverages, black coffee, unsweetened sodas, or mineral water add no calories to your diet. Lowfat milk is also a nutritious choice. Carry individually packaged sugar substitutes to sweeten beverages. Try a slice of lemon or lime in a glass of ice water.

h. Alcohol does have calories as shown below:

Table C–0–9
Alcohol calories

	Calories
American beer, 12 oz.	160
European beer, 1/2 liter	250
Cocktails, 4–6 oz. (1/2 cup)	165
Hard liquor, 1½ oz. Jigger	110
Dry wine, .25 litre (8 oz.)	200
Sweet wine, .25 liter (8 oz.)	300

C–14. Educate your appetite

a. Follow these tips to lose weight and body fat.

(1) Eat S-L-O-W-L-Y

(2) Consume less fat.

(3) Take smaller portions

(4) Consume less sugar.

(5) Take smaller bites

(6) Chew food thoroughly

(7) Eat at least three regular meals per day.

(8) Plan snacks.

b. Plan your food intake and abide by your plan. Keep a food diary and monitor your own intake.

c. Always check your intake for balanced nutrition and total calories.

d. Become aware of how many calories you're consuming- especially in snack foods. This sample calorie count shows you why:

Table C–0–10
Sample calorie chart

	Calories
Chocolate milkshake (8 oz.)	840
Coke, soda, or sugared beverage, 12 ozs.	160
French fries, 20 pieces (2" x 1/2")	275
Fruit pie (1/6th of 9-inch pie)	410
Pecan pie (1/6th of 9-inch pie)	750
Cheesecake (1/6th of 9-inch pie)	800
Ice cream, 2 scoops	200
Potato or corn chips, 1 ounce package	180
Chocolate candy, 1 ounce	150
Grapefruit or orange juice, 1 cup	100
Big Mac	557
Quarter-pounder with cheese	521
AAFES Jumbo Cheeseburger	654
Cottage cheese, creamed, 1 cup	223
Beer (European), 1/2 liter	250
Creamy salad dressing, 1/4 cup	340
Salter nuts:	
peanuts, 2 tbs	170
cashews, 25 nuts	252
mixed, 25 nuts	188

C–15. Some final advice

a. Get motivated. Forget those old excuses: "But I've got a large frame—you know, big bones," or "Everyone in my family is big—it's hereditary,"or "Gosh, I haven't weighed 180 pounds since I was 15 years old,"or "I can pass the APFT, so why lose weight?"

b. If you're serious about losing weight, it will be easy to give up these typical excuses. Dieting is up to you—if you are willing to try. Don't be angry that you have to lose weight; think of it as something you have chosen to do. You may be a good soldier now, but you want to strive to be the best.

c. Try not to be food centered. Eating should be a source of enjoyment and satisfaction. It shouldn't be a substitute for feelings of boredom, anger, loneliness, or discouragement. Occupy your time with other activities not related to food. This is important if you plan on being successful at losing weight. Make up your mind to control food instead of letting food control you.

d. Construct new habit patterns. Make your meals last longer, slow down, enjoy flavors, pause halfway through meals and don't stuff yourself. Make meals as pleasant as possible even if you are cutting down on what you eat.

e. Don't be discouraged by weight plateaus (i.e., periods when no weight loss occurs despite your dieting and exercise efforts). Your body is adjusting to your new eating habits and changes are taking place. Stick with it!

f. You are always responsible for what you are eating. Don't cheat; remember, overeating hurts no one but yourself.

g. If you need extra help with your diet contact the nutrition clinic at your nearest treatment facility. For an exercise plan the physical therapy clinic at your nearest medical treatment facility is an excellent resource; master fitness trainers are also excellent resources to consult for basic nutrition information and exercise program development.

h. Use the attached "Personal Weight Loss Progress Chart" DA Form 5511–R which is located in the back of this handbook to keep track of your progress.

i. Remember all the work it took to get you to your desired weight. Don't let this be wasted effort.

j. For best results, combine dietary plans with regular exercise to:

k. Your dining facility provides low calorie menu plans to assist in your weight control efforts (see tables C–3 and C–4 for sample menu patterns).

BE☆ALL☆YOU☆CAN BE—NUTRITIONALLY!

Table C–1
Weight Control a Personal Choice

Instead of Amount	Calories	Substitute with	Calories
1 cup sweetened applesauce	232	unsweetened applesauce	100
3 oz. beef bologna	237	lean ham	103
1 oz. natural Swiss or cheddar cheese	110	part-skim milk mozzarella	90
1 cup cream-style cottage cheese	223	low-fat (1%) cottage cheese	125
1 oz. cream cheese	106	1 oz. Neufchatel cheese	70
1 cup vanilla ice cream	257	one-half cup vanilla ice cream	129
1 cup whole milk	170	skim milk	80
3 oz. dry salami	384	canned chicken	153
1 cup sour cream	454	low-fat yogurt	123
one-half cup frozen sweetened strawberries	139	frozen unsweetened strawberries	118
one-half cup oil-pack tuna	158	water-pack tuna	126
one-half cup syrup-pack canned fruit	90	drained/rinsed canned fruit	50
2 tablespoons mayonnaise	120	2 tablespoons mustard	24
4 tablespoons regular salad dressing	300	4 tablespoons diet salad dressing	80
1 piece fruit pie	410	1 piece fresh fruit	80

Notes:
REMEMBER: A calorie saved is a calorie burned...and 3,500 calories is equal to a pound of body fat.

Table C–2
Comparison of Regular and TRIMM* meals

Regular Food Item	Calories	TRIMM Food Item	Calories
Cream of tomato soup (8 oz.)	173	Tomato bouillon (8 oz.)	36
Meat loaf (4 oz.)	227	Meat loaf (3 oz.)	170
Brown Gravy (3 oz.)	130		
O'Brien potatoes	236	Parsley potatoes (1 med. 4½ oz.) NO BUTTER	99
Club spinach (w/eggs) ½ cup	153	Spinach (w/lemon) ½ cup	29
Tossed green salad (1 cup)	10	Tossed green salad (1 cup)	10
1000 Island Dressing (2 tbsp. or 1 oz.)	165	Diet 1000 Island Dressing (2 tbsp. or 1 oz.)	54
Yellow cake, chocolate frosting (3 inch squares.)	365	Fresh fruit (1 piece)	80
Milk, whole (8 fl. oz.)	159	Milk, skim (8 fl. oz.)	80
Totals	1618		558

Notes:
* TRIMM=The Right Image of Me in the Military.
The TRIMM Sample Meal is lower in fat and sugar—thus more "nutrient dense" and lower in calories

Table C–3
Sample 1200 Calorie Menu

	Average Calories
Breakfast	
1 cup fruit or juice (unsweetened)	100
1 ounce meat, cheese or egg (prepared without added fat)	80
1 slice toast (white, whole wheat, rye, or ¾ cup dry cereal)	70
1 cup skim milk	80
Total	330
Lunch	
2-ounce serving lean meat (or 1 oz. meat and 1 oz. cheese)	160
2 slices of bread or 1 bun or roll	140
1 serving cooked vegetables (leaves and stems—25 calories per ½ cup; starchy vegetables corn, lima beans—70 calories per one-third cup)	25
1 serving tossed vegetable salad	10
Vinegar or lemon juice	—
1 piece of *small* fresh fruit (or ¼ cup drained canned fruit)	50
Total	385
Supper	
3-ounce serving lean meat (no gravy)	225
one-half cup starch (potato, rice, pasta, dry beans or starchy vegetables)	80
1 serving cooked vegetables	25
1 serving tossed vegetable salad	10
Vinegar or lemon juice	—
1 piece of *small* fresh fruit (or ¼ cup drained canned fruit)	50
1 cup skim milk	80
Total	470
Choose Zero Calorie Beverages:	
Coffee or tea without cream or sugar, club soda, mineral water, iced water with lemon or lime.	0
Approximate Total for Day	1185

Table C–4
Sample 1500—Calorie Menu

	Average Calories
Breakfast	
1 cup fruit or juice (unsweetened)	100
1 ounce meat, cheese or egg, prepared without fat	80
1 slice toast (white, whole wheat, rye)	70
One-half cup cooked cereal or ¾ cup dry cereal (unsweetened)	70
1 cup skim milk	80
1 teaspoon margarine	45
Total	445
Lunch	
3 oz. serving lean meat (or 2 oz. meat and 1 slice cheese)	210
2 slices of bread or 1 bun or 1 roll	140
1 serving cooked vegetables	25
(Leaves and stems 25 calories per ½ cup; starchy vegetables 70 calories per one-third cup)	
1 tossed vegetable salad	10
1 tablespoon low-calorie dressing	30
1 piece of fresh fruit (medium to large)	80
1 small or ¼ cup drained fruit, 40 calories	
Total	520
Dinner	
3 oz. serving lean meat	225
one-half cup starchy vegetable (with small amount added fat)	100
Serving cooked vegetable	25
1 tossed vegetable salad	10
1 tablespoon low-calorie dressing	30
1 piece fresh fruit (medium to large)	80
1 cup skim milk	80
Total	550
Choose Zero Calorie Beverages:	
Coffee or tea without cream or sugar, club soda, mineral water, ice water with lemon or lime.	0
Approximate Total for Day	1515

Glossary

Section I
Abbreviations

AD
active duty

ADS
active duty support

ADT
active duty for training

AGR
Active Guard/Reserve

APFT
Army Physical Fitness Test

ARNG
Army National Guard

ARPERCEN
US Army Reserve Personnel Center

CAR
Chief, Army Reserve

CG
commanding general

CNGB
Chief, National Guard Bureau

DA
Department of the Army

DCSPER
Deputy Chief of Staff for Personnel

DCSLOG
Deputy Chief of Staff for Logistics

DOD
Department of Defense

FORSCOM
US Army Forces Command

HQDA
Headquarters, Department of the Army

HSC
US Army Health Services Command

IRR
Individual Ready Reserve

MACOM
major Army command

RC
Reserve Component

SADT
Special Active Duty for Training

TPU
troop program unit

TRADOC
US Army Training and Doctrine Command

TSG
The Surgeon General

USAR
US Army Reserve

Section II
Terms

Body composition
The human body is composed of two major elements: lean body mass (which includes muscle, bone, and essential organ tissue) and body fat. Body fat is expressed as a percentage of total body weight that is fat. For example, an individual who weighs 200 pounds and 18 percent body fat has 36 pounds of fat. Women generally have a higher percentage of body fat than men because of genetic and hormonal differences; thus, body fat standards differ among men and women by age groups.

Health care personnel
Trained physicians (military or civilian employees or contract personnel), physician's assistants, Registered Nurses, dieticians, and physical/occupational therapists under supervision of the unit surgeon or the commander of the medical treatment facility. For the purpose of this regulation, this term includes personnel of US forces and host nations.

Overweight
A soldier is considered overweight when his or her percent body fat exceeds the standard specified in paragraph 20*c* of this regulation.

Professional military or civilian schooling
This includes all individual training courses beyond Initial Entry Training (IET). It does not include unit training involving crews and teams. IET includes basic branch course or equivalent for officers; warrant officer entry course for non-prior service personnel; and basic training, AIT, OSUT, and OST for enlisted personnel.

Satisfactory progress
Progressing toward a point to meet the body fat standards described in paragraph 19*c*. Weight loss of 3–8 pounds per month is required for satisfactory progress.

Weight control
An individual program by which each soldier attains and maintains an acceptable weight and body composition through self-motivation or involvement in an official weight control program.

Section III
Special Abbreviations and Terms
There are no special terms.

Made in the USA
Monee, IL
05 November 2026

67808344R00087